T0275680

Identification of Neural Markers Accompanying Memory

Identification of Neural Markers Accompanying Memory

Edited by

Alfredo Meneses
Department of Pharmacobiology,
CINVESTAV (Centro de Investigación y de Estudios Avanzados
del Instituto Politécnico Nacional),
Mexico

ELSEVIER

AMSTERDAM • BOSTON • HEIDELBERG • LONDON • NEW YORK • OXFORD
PARIS • SAN DIEGO • SAN FRANCISCO • SINGAPORE • SYDNEY • TOKYO

Elsevier
32 Jamestown Road, London NW1 7BY
225 Wyman Street, Waltham, MA 02451, USA

First edition 2014

Notices
Knowledge and best practice in this field are constantly changing. As new research and experience
broaden our understanding, changes in research methods, professional practices, or medical treatment
may become necessary.

Practitioners and researchers must always rely on their own experience and knowledge in evaluating and
using any information, methods, compounds, or experiments described herein. In using such information
or methods they should be mindful of their own safety and the safety of others, including parties for
whom they have a professional responsibility.

To the fullest extent of the law, neither the Publisher nor the authors, contributors, or editors, assume
any liability for any injury and/or damage to persons or property as a matter of products liability,
negligence or otherwise, or from any use or operation of any methods, products, instructions, or ideas
contained in the material herein.

British Library Cataloguing-in-Publication Data
A catalogue record for this book is available from the British Library

Library of Congress Cataloging-in-Publication Data
A catalog record for this book is available from the Library of Congress

ISBN: 978-0-12-408139-0

For information on all Elsevier publications
visit our website at store.elsevier.com

This book has been manufactured using Print On Demand technology. Each copy is produced to order
and is limited to black ink. The online version of this book will show color figures where appropriate.

Working together
to grow libraries in
developing countries

www.elsevier.com • www.bookaid.org

"To my wonderful wife Erika, my daughter Sofia and my two little Angel"

Contents

List of Contributors   xi

1 Introduction   1
Alfredo Meneses

Neurotransmitters and Memory: Introduction   1
References   2

2 Neurotransmitters and Memory: Cholinergic, Glutamatergic, GaBAergic, Dopaminergic, Serotonergic, Signaling, and Memory   **5**
Alfredo Meneses

Neurotransmitters and Memory   5
    Cholinergic   5
    Glutamate   7
    GABA   10
    Dopamine   11
    Serotonin   13
5-HT Systems and Neurobiological Markers Related to Memory Systems   14
5-HT Neural Markers and Memory   14
Brain Areas, Biochemical Pathways, Cognitive-Enhancing Effects of 5-HT Receptor Drugs   16
5-HT Pathways, Receptors, and Transporter: Memory Functions and Dysfunctions   17
Protocols of Training/Testing, Memory Tasks, and Drugs   18
Loci, Mechanisms of Action, and Memory Tasks   21
Memory Tasks and Signaling   21
Brain Areas, Neurotransmitters Systems, Drugs: Cognitive and Behavioral Demand of Memory Tasks and Protocols of Training: a Final Consideration   23
Signaling and Memory   23
    Memories and Molecular Traces   25

Acknowledgments   30
References   30

3   The Role of GABA in Memory Processes   47
Antonella Gasbarri and Assunta Pompili

Introduction   47
GABA Receptors   48
GABA and Its Correlation with Memory   51
    GABA, Memory and BZ   52
    GABA and Spatial Memory   53
    GABA and Fear Memory   55
References   56

4   Involvement of Glutamate in Learning and Memory   63
Antonella Gasbarri and Assunta Pompili

Introduction   63
Glutamate Receptors   64
    Ionotropic Receptors of Glutamate   64
    Metabotropic Receptors of Glutamate   67
Glutamate, Memory, and HF   68
    Glutamate and Addiction   69
    Glutamatergic System and Nervous Diseases   70
Conclusions   72
References   72

5   Dopamine and Memory   79
Ryan T. LaLumiere

Pharmacology of Dopamine   79
Anatomy of the Dopaminergic System   80
Basal Ganglia   82
Working Memory   83
Reinforcement Learning   84
Dopamine and Neural Plasticity   85
Dopamine and Memory Consolidation   86
Conclusions   88
References   89

**6 Unpacking Memory Processes: Using the Attribute Model
    to Design Optimal Memory Tests for Rodent Models**   **95**
Michael R. Hunsaker

Introduction   **95**
Attribute Model   **96**
   Memory Systems   **96**
   Specific Attributes   **96**
   Processes Associated with Each Attribute   **97**
   Attributes Map onto Neural Substrates   **98**
   Interactions Among Attributes   **99**
Applying the Attribute Model   **101**
   General Advice   **101**
   Specific Application of the Attribute Model   **105**
Conclusions   **107**
References   **109**

7 Protein Synthesis and Memory: A Word of Caution   **112**
*Roberto Agustín Prado-Alcalá, Andrea C. Medina, Norma Serafín
and Gina L. Quirarte*

Acknowledgments   **117**
References   **117**

**8 Basic Elements of Signal Transduction Pathways Involved in
    Chemical Neurotransmission**   **121**
Claudia González-Espinosa and Fabiola Guzmán-Mejía

Introduction   **121**
Some Central Concepts on Cell-to-Cell Communication   **122**
   Neurotransmitters   **123**
G-Protein-Coupled Receptors   **124**
GPCR Activation   **125**
G-Proteins   **125**
Receptor Desensitization   **125**
β-Arrestin-Dependent Signaling   **126**
Small GTPases   **127**
The Second Messengers   **127**
Cyclic Adenosine Monophosphate   **127**
Inositol 1,4,5 Triphosphate   **128**

Calcium   **128**
Mitogen-Activated Protein Kinases   **129**
The Transcription Factors   **130**
Epigenetic Modifications   **131**
References   **131**

**9 A Role for Learning and Memory in the Expression
    of an Innate Behavior: The Case of Copulatory Behavior   135**
    Gabriela Rodríguez-Manzo and Ana Canseco-Alba

Introduction   **135**
Male Rat Sexual Behavior   **136**
Effects of Sexual Experience on Copulatory Behavior Expression   **137**
Effects of Sexual Experience on Brain Functioning   **140**
Brain Regions Involved in Sexual Experience-Induced
Behavioral Changes   **141**
Conclusions   **142**
References   **143**

10 Memory Disorders: The Diabetes Case   149
    Gustavo Liy-Salmeron and Shely Azrad

Introduction   **149**
Glucose Regulation, Diabetes, and AD   **150**
Insulin, IR, and AD   **150**
Apolipoprotein E and AD   **151**
Cholesterol and AD   **152**
Caffeine and AD   **154**
Conclusions   **154**
References   **155**

List of Contributors

Roberto Agustín Prado-Alcalá Departamento de Neurobiología Conductual y Cognitiva, Instituto de Neurobiología, Universidad Nacional Autónoma de México, Campus Juriquilla, Querétaro, Qro 76230, México

Shely Azrad Facultad de Ciencias de la Salud, Universidad Anáhuac México Norte, Huixquilucan Edo. de México

Ana Canseco-Alba Pharmacobiology Department, Center for Research and Advanced Studies (Cinvestav), South Campus, Mexico City, Mexico

Antonella Gasbarri Department of Applied Clinical and Biotechnologic Sciences, University of L'Aquila, Italy

Claudia González-Espinosa Pharmacobiology Department, Center for Research and Advanced Studies (Cinvestav), South Campus, Mexico City, México

Fabiola Guzmán-Mejía Pharmacobiology Department, Center for Research and Advanced Studies (Cinvestav), South Campus, Mexico City, México

Michael R. Hunsaker Department of Psychiatry and Behavioral Sciences and MIND Institute, University of California, Davis Medical Center, Sacramento, CA

Ryan T. LaLumiere Department of Psychology, University of Iowa, Iowa City, IA 52242

Gustavo Liy-Salmeron Facultad de Ciencias de la Salud, Universidad Anáhuac México Norte, Huixquilucan Edo. de México

Andrea C. Medina Departamento de Neurobiología Conductual y Cognitiva, Instituto de Neurobiología, Universidad Nacional Autónoma de México, Campus Juriquilla, Querétaro, Qro 76230, México

Alfredo Meneses Department of Pharmacobiology, CINVESTAV, México

Assunta Pompili Department of Applied Clinical and Biotechnologic Sciences, University of L'Aquila, Italy

Gina L. Quirarte Departamento de Neurobiología Conductual y Cognitiva, Instituto de Neurobiología, Universidad Nacional Autónoma de México, Campus Juriquilla, Querétaro, Qro 76230, México

Gabriela Rodríguez-Manzo Pharmacobiology Department, Center for Research and Advanced Studies (Cinvestav), South Campus, Mexico City, Mexico

Norma Serafín Departamento de Neurobiología Conductual y Cognitiva, Instituto de Neurobiología, Universidad Nacional Autónoma de México, Campus Juriquilla, Querétaro, Qro 76230, México

1 Introduction

Alfredo Meneses

Department of Pharmacobiology, CINVESTAV, México

Neurotransmitters and Memory: Introduction

Memory is a basic function of the brain and fundamental in our life. Memory may be defined according to its content, in relation to time and its neurobiological basis: in the former case, as declarative/explicit or nondeclarative/implicit memory, and regarding time, as short-term memory (STM) or working and long-term memory (LTM) (Davis and Squire, 1984; Izquierdo et al., 1999, 2006); the latter depends on protein and mRNA synthesis (Meneses et al., 2011). Considering that memory is a field of scientific investigation in constant expansion, hence in this book the aim is offering a brief and introductory overview for students of the area, using relevant and (and when is possible) recent references. Diverse brain areas (Squire and Zola, 1996; Eichenbaum, 2013), neurotransmitters (see below), and cell signaling have been associated to memory (Vianna et al., 2000) and its alterations.

Firstly, extensive evidence indicates that disruption of cholinergic function is characteristic of aging and Alzheimer's disease (AD), and experimental manipulation of the cholinergic system in laboratory animals suggests age-related cholinergic dysfunction may play an important role in cognitive deterioration associated with aging and AD (Decker and McGaugh, 2004; McGaugh, 1989; Myhrer, 2003). Recent investigation, however, suggests that cholinergic dysfunction does not provide a complete account of age-related cognitive deficits and that age-related changes in cholinergic function typically occur within the context of changes in several other neuromodulatory systems. Interactions between the cholinergic system and several of other neurotransmitters and neuromodulators (including norepinephrine, dopamine, serotonin, GABA, opioid peptides, galanin, substance P, and angiotensin II) may be important in learning and memory (Decker and McGaugh, 2004; Reiss et al., 2009). It is important to consider not only the independent contributions of age-related changes in neuromodulatory systems to cognitive decline, but also the contribution of interactions between neurotransmission systems to the learning and memory deficits associated with aging and AD (Decker and McGaugh, 2004).

In perspective, if receptors for all agents (e.g., hormones, trophic factors odorants, peptides) in addition to the transmitters were counted, a total of 1000 would not be surprising (Cooper et al., 2003). Before this abundance of neurotransmitters,

Identification of Neural Markers Accompanying Memory. DOI: http://dx.doi.org/10.1016/B978-0-12-408139-0.00001-8

then, among the neurotransmitters mentioned in this chapter, we are including the cholinergic (Bentley et al., 2011; Graef et al., 2011), glutamatergic (Piers et al., 2012; chapter 3 this volume), GABAergic (Reiss et al., 2009; chapter 4 this volume), dopaminergic (Reiss et al., 2009; chapter 5 this volume), and serotonergic (serotonin, 5-hydroxytryptamine or 5-HT) (Altman and Normile, 1988; Ogren, 1985; Rodríguez et al., 2012; see below). Importantly, the behavioral endophenotypes had become an important tool (chapter 6 this volume). An excellent, fresh, and new advance is about a role for learning and memory in the expression of an innate behavior (chapter 9 this volume). Of course, protein synthesis and memory had become an important subject (chapter 7 this volume).

As a detailed description of the neurotransmission systems included herein is beyond the present work, then a shallow overview is provided. An excellent atlas about neuroactive substances, their projections and receptors is available (Tohyama et al., 1998); hence, in this work the contribution of pharmacology to research on the mechanisms of memory formation (McGaugh and Izquierdo, 2000) is one of the major focus. Certainly, the transporters of neurotransmission systems play a crucial function in the regulation of intra-synaptic concentrations and play an important role in memory formation, amnesia, anti-amnesic effects, and forgetting (see e.g., Tellez et al., 2012a,b). Likewise, hippocampal neurogenesis and forgetting are important (Frankland et al., 2013). Finally, the diabetes case (chapter 10 this volume) allows illustrating the variety of memory disorders. The basic elements of signal transduction pathways are provided (chapter 8 this volume).

References

Altman, H.J., Normile, H.J., 1988. What is the nature of the role of the serotonergic nervous system in learning and memory: prospects for development of an effective treatment strategy for senile dementia. Neurobiol. Aging. 9 (5–6), 627–638.

Bentley, P., Driver, J., Dolan, R.J., 2011. Cholinergic modulation of cognition: insights from human pharmacological functional neuroimaging. Prog. Neurobiol. 94 (4), 360–388.

Cooper, J.R., Bloom, F.E., Roth, R.H., 2003. The Biochemical Basis of Neuropharmacology. Oxford University Press, New York, NY.

Davis, H.P., Squire, L.R., 1984. Protein synthesis and memory: a review. Psychol. Bull. 96 (3), 518–559.

Decker, M.W., McGaugh, J.L., 2004. The role of interactions between the cholinergic system and other neuromodulatory systems in learning and memory. Synapse. 7 (2), 151–168.

Eichenbaum, H., 2013. What H.M. taught us. J. Cogn. Neurosci. 25 (1), 14–21.

Frankland, P.W., Köhler, S., Josselyn, S.A., 2013. Hippocampal neurogenesis and forgetting. Trends Neurosci. 36 (9), 497–503.

Graef, S., Schönknecht, P., Sabri, O., Hegerl, U., 2011. Cholinergic receptor subtypes and their role in cognition, emotion, and vigilance control: an overview of preclinical and clinical findings. Psychopharmacology (Berl.). 215 (2), 205–229.

Izquierdo, I., Medina, J.H., Vianna, M.R., Izquierdo, L.A., Barros, D.M., 1999. Separate mechanisms for short- and long-term memory. Behav. Brain Res. 103 (1), 1–11.

Izquierdo, I., Bevilaqua, L.R., Rossato, J.I., Bonini, J.S., Medina, J.H., Cammarota, M., 2006. Different molecular cascades in different sites of the brain control memory consolidation. Trends Neurosci. 29 (9), 496–505.

McGaugh, J.L., 1989. Dissociating learning and performance: drug and hormone enhancement of memory storage. Brain Res. Bull. 23 (4–5), 339–345.

McGaugh, J.L., Izquierdo, I., 2000. The contribution of pharmacology to research on the mechanisms of memory formation. Trends Pharmacol. Sci. 21 (6), 208–210.

Meneses, A., Pérez-García, G., Ponce-Lopez, T., Castillo, C., 2011. 5-HT$_6$ receptor memory and amnesia: behavioral pharmacology—learning and memory processes. Int. Rev. Neurobiol. 96, 27–47.

Myhrer, T., 2003. Neurotransmitter systems involved in learning and memory in the rat: a meta-analysis based on studies of four behavioral tasks. Brain Res. Rev. 41, 268–287.

Ogren, S.O., 1985. Evidence for a role of brain serotonergic neurotransmission in avoidance learning. Acta Physiol. Scand. Suppl. 544, 1–71.

Piers, T.M., Kim, D.H., Kim, B.C., Regan, P., Whitcomb, D.J., Cho, K., 2012. Translational concepts of mGluR5 in synaptic diseases of the brain. Front. Pharmacol. 3, 199.

Reis, H.J., Guatimosim, C., Paquet, M., Santos, M., Ribeiro, F.M., Kummer, A., et al., 2009. Neuro-transmitters in the central nervous system & their implication in learning and memory processes. Curr. Med. Chem. 16 (7), 796–840.

Rodríguez, J.J., Noristani, H.N., Verkhratsky, A., 2012. The serotonergic system in ageing and Alzheimer's disease. Prog. Neurobiol. 99, 15–41.

Squire, L.R., Zola, S.M., 1996. Structure and function of declarative and nondeclarative memory systems. Proc. Natl. Acad. Sci. USA. 93 (24), 13515–13522.

Tellez, R., Gómez-Víquez, L., Meneses, A., 2012a. GABA, glutamate, dopamine and serotonin transporters expression on memory formation and amnesia. Neurobiol. Learn. Mem. 97 (2), 189–201.

Tellez, R., Gómez-Viquez, L., Liy-Salmeron, G., Meneses, A., 2012b. GABA, glutamate, dopamine and serotonin transporters expression on forgetting. Neurobiol. Learn. Mem. 98 (1), 66–77.

Tohyama, M., Takatsuji, K., Kantha, S.S. (Eds.), 1998. Atlas of Neuroactive Substances and Their Receptors in the Rat. Oxford University Press, Oxford.

Vianna, M.R., Izquierdo, L.A., Barros, D.M., Walz, R., Medina, J.H., Izquierdo, I., 2000. Short- and long-term memory: differential involvement of neurotransmitter systems and signal transduction cascades. An. Acad. Bras. Cienc. 72 (3), 353–364.

2 Neurotransmitters and Memory: Cholinergic, Glutamatergic, GaBAergic, Dopaminergic, Serotonergic, Signaling, and Memory

Alfredo Meneses

Department of Pharmacobiology, CINVESTAV, México

Neurotransmitters and Memory

Cholinergic

Terry and Buccafusco in 2003 revised the cholinergic hypothesis about memory, which was initially presented over 20 years ago, indicating that a dysfunction of acetylcholine containing neurons in the brain contributes substantially to the cognitive decline observed in those with advanced age and Alzheimer's disease (AD). This premise has since served as the basis for the majority of treatment strategies and drug development approaches for AD to date (Terry and Buccafusco, 2003; see also Bentley et al., 2011; Cummings et al., 2012; Decker and McGaugh, 2004). Studies of the brains of patients who had mild cognitive impairment (MCI) or early stage AD in which choline acetyltransferase and/or acetylcholinesterase activity was unaffected (or even upregulated) have, however, led some to challenge the validity of the hypothesis as well as the rationale for using cholinomimetics to treat the disorder, particularly in the earlier stages (Terry and Buccafusco, 2003). These challenges primarily are based on assays of postmortem enzyme activity, which should be taken in perspective and evaluated within the wide range of cholinergic abnormalities known to exist in both aging and AD (e.g., Zhang et al., 2012; Zhou et al., 2012). In addition, the results of both postmortem and antemortem studies in aged humans and AD patients, as well as preclinical animal experimentation, suggest that a host of cholinergic abnormalities including alterations in choline transport, acetylcholine release, nicotinic and muscarinic receptor expression, neurotrophin support, and perhaps axonal transport may all contribute to cognitive abnormalities in aging and AD (Terry and Buccafusco, 2003; see also Graef et al., 2011; Martyn et al., 2012) as well as other neurotransmitter

Identification of Neural Markers Accompanying Memory. DOI: http://dx.doi.org/10.1016/B978-0-12-408139-0.00002-X

systems. Cholinergic abnormalities may also contribute to noncognitive behavioral abnormalities as well as the deposition of toxic neuritic plaques in AD. Therefore, cholinergic-based strategies will likely remain valid approaches to rational drug development for the treatment of AD and other forms of dementia (Terry and Buccafusco, 2003). Focus on the cholinergic system in aging and neuronal degeneration reveals that the basal forebrain cholinergic complex comprising medial septum, horizontal and vertical diagonal band of Broca, and nucleus basalis of Meynert provides the mayor cholinergic projections to the cerebral cortex and hippocampus (Schliebs and Arendt, 2011). The cholinergic neurons of this complex have been assumed to undergo moderate degenerative changes during aging, resulting in cholinergic hypofunction that has been related to the progressing memory deficits with aging. However, the previous view of significant cholinergic cell loss during aging has been challenged (Schliebs and Arendt, 2011). Neuronal cell loss was found predominantly in pathological aging, such as AD, while normal aging is accompanied by a gradual loss of cholinergic function caused by dendritic, synaptic, and axonal degeneration as well as a decrease in trophic support (Schliebs and Arendt, 2011; see also Tremblay et al., 2010). As a consequence, decrements in gene expression, impairments in intracellular signaling, and cytoskeletal transport may mediate cholinergic cell atrophy, which finally leads to the known age-related functional decline in the brain including aging-associated cognitive impairments (Schliebs and Arendt, 2011). However, in pathological situations associated with cognitive deficits, such as Parkinsons's disease, Down syndrome, progressive supranuclear palsy, Creutzfeldt–Jakob disease, Korsakoff's syndrome, traumatic brain injury, significant degenerations of basal forebrain cholinergic cells, have been observed. In presenile (early onset), and in the advanced stages of late-onset AD, a severe loss of cortical cholinergic innervation has extensively been documented. In contrast, in patients with MCI (a prodromal stage of AD), and early forms of AD, apparently no cholinergic neurodegeneration but a loss of cholinergic function occurs (Schliebs and Arendt, 2011). In particular imbalances in the expression of NGF, its precursor proNGF, the high and low NGF receptors, trkA and p75NTR, respectively, changes in acetylcholine release, high-affinity choline uptake, as well as alterations in muscarinic and nicotinic acetylcholine receptor expression, may contribute to the cholinergic dysfunction (Reiss et al., 2009; Schliebs and Arendt, 2011). These observations support the suggestion of a key role of the cholinergic system in the functional processes that lead to AD. Malfunction of the cholinergic system may be tackled pharmacologically by intervening in cholinergic as well as neurotrophic signaling cascades that have been shown to ameliorate the cholinergic deficit at early stages of the disease and slow down the progression (Schliebs and Arendt, 2011). However, in contrast to many other disorders demential ones, in AD the cholinergic dysfunctions are accompanied by the occurrence of two major histopathological hallmarks such as β-amyloid plaques and neurofibrillary tangles, provoking the question whether they play a particular role in inducing or mediating cholinergic dysfunction in AD (Schliebs and Arendt, 2011). Indeed, there is abundant evidence that β-amyloid may trigger cholinergic dysfunction through action on α7-nicotinic acetylcholine receptors, affecting NGF signaling, mediating tau phosphorylation, interacting with acetylcholinesterase, and specifically affecting the proteome in cholinergic neurons (Schliebs and Arendt, 2011). Therefore, an early onset of an anti-β-amyloid strategy may additionally be potential in preventing aging-associated cholinergic deficits and cognitive impairments (Schliebs and Arendt, 2011). Herein, it is important to highlight that AD likely represents the most known psychiatry disorder where memory is (progressively) impaired; however, others (e.g., depression, Parkinson's disease; see Millan et al., 2012) also present memory deficits. Certainly, beyond the implication of cholinergic system in memory and therapeutic applications, the muscarinic antagonist scopolamine had been a remarkable amnesic agent (Klinkenberg and Blokland, 2010).

In addition, therapeutic options for AD are currently limited to symptomatic treatment that only provides modest and temporary maintenance of cognitive and memory functions, without altering disease progression (Haas, 2012). Although a variety of therapeutics targeting amyloid production or plaque degradation as well as tau hyperphosphorylation and aggregation have been proposed, examined in preclinical models and introduced into clinical trials; many have failed to provide significant therapeutic benefit (Haas, 2012). Concerns over the adequacy of currently used preclinical models, in addition to questions pertaining to the timing of therapeutic administration, regarding synaptic and neuronal loss have been raised and are further complicated by the genetic diversity of individual patients (Haas, 2012; for further complications with dysfunctional memory see also Eichenbaum, 2013). Haas (2012) offers a brief overview of AD pathophysiology and the currently approved therapeutics, with focus on therapeutics currently evaluated in preclinical models and clinical trials. Probably, other complications are the intricate and complex problems of memory, and the little are known of how the brain works.

Glutamate

Another neurotransmission system traditionally associated to memory is the glutamatergic system (Morris, 2013; Niciu et al., 2012; Radley et al., 2007; Reis et al., 2009). Glutamate receptors may be divided into two broad categorizations: ionotropic and metabotropic receptors (Niciu et al., 2012); the ionotropic receptors are subdivided into N-methyl-D-aspartate (NMDA) and alpha-amino-3-hydroxy-5-methyl-4-isoxazole-propionic acid (AMPA)/kainite, while metabotropic receptors are grouped (mGluR1, mGluR5, etc.) (Niciu et al., 2012). The glutamatergic ionotropic receptors have a role in functions in the central nervous system, including regulating neurodevelopment and synaptic plasticity, learning and memory, and excitotoxicity (Gonda, 2012). Due to their complex involvement in the above processes, NMDA receptors (NMDARs) have been established to play a role in the etiopathology of several neuropsychiatric disorders, including ischemia and traumatic brain injury, neurodegenerative disorders, and schizophrenia. NMDARs contain multiple types of subunits with distinct functional and pharmacological properties making the picture more complex (Gonda, 2012). These receptors also offer multiple target-binding sites for drugs; however, early broad-spectrum NMDAR antagonists had limited clinical use due to their intolerable adverse effect profile. The discovery of several types of subunit-selective NMDAR antagonists may offer valuable therapeutic possibilities for several disorders, with improved clinical efficacy and decreased side effects. However, in spite of our increasing knowledge concerning the involvement of NMDARs in pathological processes, molecules with a selective action, tolerable side effect profile, and good clinical efficacy are still only in clinical development in the majority of cases. Certainly, as already observed NMDARs offer a novel opportunity in the treatment of various neuropsychiatric conditions (Gonda, 2012). On the other hand, AMPA receptors (AMPARs) mediate the majority of fast excitatory synaptic transmission in the brain. Dynamic changes in neuronal synaptic efficacy, termed synaptic plasticity, are thought to underlie information coding and storage in learning and memory (Anggono and Huganir, 2012). One major mechanism that regulates synaptic strength involves the tightly regulated trafficking of AMPARs into and out of synapses. The life cycle of AMPARs from their biosynthesis, membrane trafficking, and synaptic targeting to their degradation is controlled by a series of orchestrated interactions with numerous intracellular regulatory proteins. The regulation of AMPAR trafficking, focusing on the roles of several key intracellular AMPAR interacting proteins (Anggono and Huganir, 2012; see also Lynch, 2006). Lynch (2006) informs us that ampakines are a structurally diverse family of small molecules that positively modulate AMPA-type glutamate receptors, and thereby enhance fast, excitatory transmission throughout the brain. Surprisingly, ampakines have discrete effects on brain activity and behavior. Because their excitatory synaptic targets mediate communication between cortical regions, serve as sites of memory encoding, and regulate the production of growth factors, ampakines have a broad range of potential therapeutic applications. Several of these possibilities have been tested with positive results in preclinical models; preliminary clinical work has also been encouraging (Lynch, 2006).

Jane et al. (2009) highlight that compared to the other glutamate receptors, progress in the understanding of the functions of kainate receptors (KARs) has lagged behind, due mainly to the relative lack of specific pharmacological tools. Over the last decade, subunit-selective agonists (e.g., ATPA and 5-iodowillardiine) and orthosteric (e.g., LY382884 and ACET) and allosteric antagonists for KARs that contain GluK1 (GluR5) subunits have been developed. However, no selective ligands for the other KAR subunits have been identified (Jane et al., 2009). The use of GluK1 antagonists has enabled several functions of KARs to be identified, containing this subunit. Thus, KARs had been shown to regulate excitatory and inhibitory synaptic transmission. In the case of the regulation of L-glutamate release, they can function as facilitatory autoreceptors or inhibitory autoreceptors during repetitive synaptic activation and can respond to ambient levels of L-glutamate to provide a tonic regulation of L-glutamate release. KARs also contribute with a component of excitatory synaptic transmission at certain synapses. They can also act as triggers for both physiological models long-term potentiation (LTP) and long-term depression (LTD) and rapid alterations in their trafficking can result in altered synaptic transmission during both synaptic plasticity and neuronal development (Jane et al., 2009). It should be noted that while LTP had been proposed as a model for memory study, LTD has been proposed as a model for forgetting, but Kemp and Manahan-Vaughan (2007) propose that LTD contributes directly to hippocampal information storage; thus, LTP and LTD enable distinct and separate forms of information storage, which together facilitate the generation of a spatial cognitive map. Returning to the KARs, which contribute to synchronized rhythmic activity in the brain and are involved in forms of learning and memory (Jane et al., 2006). Regarding therapeutic indications, antagonists for GluK1 have shown positive activity in animal models of pain, migraine, epilepsy, stroke, and anxiety. This potential has now been confirmed in dental pain and migraine in initial studies in man (Jane et al., 2006).

Finally, Morris (2013) offers us a personal point of view indicating that the glutamate antagonist R-2-amino-5-phosphonopentanoate (d-AP5) blocked the induction of LTP, *in vitro* (in hippocampal area CA3 to CA1) without apparent effect on baseline synaptic transmission. This dissociation was one of the key triggers for an explosion of interest in glutamate receptors, and much has been discovered since that collectively contributes to our contemporary understanding of glutamatergic synapses (Morris, 2013). The NMDARs participate in memory encoding in the hippocampus, visual cortical plasticity, sensitization in pain, and other functions (Morris, 2013). Morris (2013) also highlights that the NMDARs activation is essential for memory encoding, though not for storage; a notion that took time to develop and to be accepted. Along the way, there have been confusions, challenges, and surprises surrounding the idea that activation of NMDARs can trigger memory (Morris, 2013), discussing some new directions of interest with respect to the functional role of the NMDAR in cognition and LTP. Collingridge et al. (2013) highlight that the NMDARs as a target for cognitive enhancement play an important role in neural plasticity, learning and memory; including LTP and LTD. As cognitive decline is a major problem facing an aging human population, so much so that its reversal has become an important goal for scientific research and pharmaceutical development, and enhancement of NMDAR function is a core strategy toward this goal (Collingridge et al., 2013). Collingridge et al. (2013) indicate some of the major ways of potentiating NMDAR function by both direct and indirect modulation; mentioning that both positive and negative modulation can enhance function suggesting that a subtle approach correcting imbalances in particular clinical situations will be required. Excessive activation and the resultant deleterious effects will need to be carefully avoided (Collingridge et al., 2013).

Certainly, activity-dependent plasticity of glutamatergic synapses, such as LTP and LTD, gained center stage in the study of learning, memory, and experience-dependent refinement of neural circuits (Maffei, 2011). Both LTP and LTD are extensively studied and their relevance to brain function is widely accepted, new experimental and theoretical work recently demonstrates that brain development and function relies on additional forms of plasticity, some of which occur at non-glutamatergic synapses (Maffei, 2011). For instance, the strength of GABAergic synapses is modulated by activity, and new functions for inhibitory synaptic plasticity are emerging (Maffei, 2011). Together with excitatory neurons, inhibitory neurons shape the excitability and dynamic range of neural circuits. Thus, the understanding of inhibitory synaptic plasticity is crucial to fully comprehend the physiology of brain circuits. Plasticity at GABAergic synapses may contribute to circuit function (Maffei, 2011).

GABA

The gamma-aminobutyric acid (GABA) receptors respond to the neurotransmitter GABA, the chief inhibitory neurotransmitter in the vertebrate central nervous system. The ionotropic receptors of GABA are the $GABA_A$ receptors and are ligand-gated ion channels, while the metabotropic are the $GABA_B$ receptors (Cooper et al., 2003; Olsen and Sieghart, 2008). A subclass of ionotropic GABA receptors, insensitive to typical allosteric modulators of $GABA_A$ receptor channels such as benzodiazepines and barbiturates, was designated $GABA_C$ (Farrant, 2001). Pharmacology evidence had provided strong support for the GABA involvement in memory. For instance, some antiepileptic (GABAergic) dugs seem produce modest memory deficits (e.g., barbiturates, benzodiazepines), the search for new antiepileptic drugs that will improve, or at least not impair, cognitive functions (Czubak et al., 2010).

Herein, an important parenthesis, we should note that the timing of drug treatment is crucial; for instance, posttraining treatments thus provide a means of distinguishing drug effects on memory from other effects on performance, as the subjects can be drug-free during both acquisition and retention testing (Roozendaal and McGaugh, 2011). Moreover, pretreatment with benzodiazepines like diazepam impairs memory. For instance, the immediate posttraining administration of the GABA antagonist, bicuculline, or of the Cl-channel blockers, picrotoxin enhances memory (Izquierdo et al., 1990). These drugs are effective when injected into the amygdaloid nucleus. Intraamygdala muscimol has an opposite effect. Notably, all this suggests that memory is modulated at the posttraining period by $GABA_A$ receptors. The pre-, but not posttraining systemic administration of benzodiazepines hinders, and that of inverse agonists, or of the benzodiazepine antagonist, flumazenil enhances retention of diverse tasks (Izquierdo et al., 1990). Flumazenil, at doses lower than those that cause an enhancement, antagonizes the effect of benzodiazepine agonists and inverse agonists. This suggests that memory is modulated during acquisition by endogenous benzodiazepine receptor ligands: possibly the diazepam discovered in brain (Izquierdo et al., 1990). Pretraining intraamygdala muscimol administration depresses memory, at doses several times higher than those that are effective posttraining. Pretraining injection of Ro 5-4864 has no effect. This suggests that the release of endogenous benzodiazepines during training may modulate a $GABA_A$ receptor complex, possibly in the amygdala, making it more sensitive to muscimol or Ro 5-4864 in the immediate posttraining period (Izquierdo et al., 1990).

Certainly, benzodiazepine site agonists (e.g., diazepam) are well known to impair cognition, inasmuch as benzodiazepines exert their effects via modulation of $\alpha 1$-, $\alpha 2$-, $\alpha 3$-, and $\alpha 5$-containing $GABA_A$ receptors, the cognition-impairing effects of diazepam must be associated with one or several of these subtypes (Atack, 2011). Of these different subtypes, $\alpha 5$-containing $GABA_A$ receptors represent an attractive option as the "cognition" subtype based upon the preferential localization of these receptors within the hippocampus and the well-established role of the hippocampus in learning and memory. It is hypothesized that an inverse agonist selective for the $\alpha 5$ subtype should enhance cognition (Atack, 2011). For example, L-655708, a partial inverse agonist with 50- to 100-fold higher affinity for the $\alpha 5$ relative to the $\alpha 1$, $\alpha 2$, and $\alpha 3$ subtypes of GABA$_A$ receptors, enhanced cognitive performance in rats. Unfortunately, however, pharmacokinetic properties of this compound prevented it being developed further. Atack (2011) provides us with an important and illustrating example. The success of the selective efficacy approach on the $\alpha 2/\alpha 3$-selective agonist project led a similar paradigm being adopted for the $\alpha 5$ project. The starting point for this strategy was the triazolopyridazine 3 which, like MRK-536, possessed a degree of both $\alpha 5$ binding- and efficacy-selectivity. By changing the core from a triazolopyridazine to a triazolophthalazine structure, $\alpha 5$-binding selectivity was lost but with subsequent optimization, compounds with the desired profile (low or antagonist efficacy at the $\alpha 1$, $\alpha 2$, and $\alpha 3$ subtypes and marked inverse agonism at $\alpha 5$-containing receptors) could be achieved, allowing the clinical candidate $\alpha 5$IA as well as the structurally related pharmacological tool compound $\alpha 5$IA-II to be identified (Atack, 2011). Finally, a degree of $\alpha 5$-efficacy selectivity was achieved the pyridazine series but metabolic instability within this chemotype limited its further optimization. Overall, evidence revised by Atack (2011) demonstrates the feasibility of adopting a selective efficacy approach in the identification of $\alpha 5$-selective GABA$_A$ receptor inverse agonists. In contrast, Koh (2012) highlights that amnesic MCI (aMCI) is associated with increased activation in the hippocampal CA3-dentate region. Excess CA3 activity also occurs in aged rats with memory impairment. Therapies to counter such excess activity might include antiepileptics or agonists for GABA$_A$ $\alpha 5$ receptors, which regulate tonic inhibition. Use of GABA$_A$ $\alpha 5$ agonists may seem unexpected because GABA$_A$ $\alpha 5$ inverse agonists were developed as cognitive enhancers. Apparently, inverse agonists, while yielding benefit in normal young adult rats, are not effective in treating memory loss in aged rats (Koh, 2012). Instead, aged rats showed improved memory after treatment with selective GABA$_A$ $\alpha 5$ agonists and with certain antiepileptics. These benefits of treatment are consistent with the concept that excess activity in the CA3 of the hippocampus is a dysfunctional condition contributing to age-associated memory impairment. Because excess hippocampal activation is also observed in MCI further findings support the use of antiepileptic or GABA$_A$ $\alpha 5$ agonist therapy in aMCI. Such therapy, in addition to memory improvement, may also have disease-modifying potential because hippocampal overactivity in aging/MCI predicts further cognitive decline and conversion to AD (Koh, 2012).

Dopamine

The dopaminergic system plays also a prominent role in memory (Beaulieu and Gainetdinov, 2011). Thus, G-protein-coupled dopamine (DA) receptors (D1, D2, D3, D4, and D5) mediate all of the physiological functions of the catecholaminergic neurotransmitter DA, ranging from voluntary movement and reward to hormonal regulation and hypertension. Pharmacological agents targeting dopaminergic neurotransmission have been clinically used in the management of several neurological and psychiatric disorders (e.g., Parkinson's disease, schizophrenia, bipolar disorder, Huntington's disease, attention deficit hyperactivity disorder (ADHD), and Tourette's syndrome). For instance, the role of DA D1 receptors in prefrontal cortex (PFC) function, including working memory, is well acknowledged (Takahashi et al., 2012; see also Chapter 5). According with Floresco (2013), studies on PFC of DA function have revealed its essential role in mediating a variety of cognitive and executive functions. A general principle that has emerged (mainly studies on working memory) is that PFC DA, acting via D1 receptors, regulates cognition in accordance to an "inverted-U"-shaped function, so that too little or too much activity has detrimental effects on performance (Floresco, 2013). However, contemporary studies have indicated that the receptor mechanisms through which mesocortical DA regulates different aspects of behavioral flexibility can vary considerably across different DA receptors and cognitive operations (Floresco, 2013). Thus, set shifting is dependent on a cooperative interaction between PFC D1 and D2 receptors, yet, supranormal stimulation of these receptors does not appear to have detrimental effects on this function. On the other hand, modification of cost–benefit decision biases in situations involving reward uncertainty is regulated in complex and sometimes opposing ways by PFC D1 versus D2 receptors (Floresco, 2013). Collectively findings suggest that the "inverted-U"-shaped dose–response curve underlying D1 receptor modulation of working memory is not a one-size-fits-all function (Floresco, 2013). Rather, it appears that mesocortical DA exerts its effects via a family of functions, wherein reduced or excessive DA activity can have a variety of effects across different cognitive domains (Floresco, 2013). Moreover, it should be noted preclinical investigation indicates that post-training administration of d-amphetamine (likely acting via DA system; Wardle et al. (2012); for review see McGaugh (1973)) improved memory consolidation (Oscos et al., 1988) but lacking its effect in intracerebral infusions immediately after each conditioning session (Dalley et al., 2005). Notably, apathy associated with AD had been tested with dextroamphetamine challenge (Lanctôt et al., 2008). In addition, impairment of language function (aphasia) is one of the most common neurological symptoms (Bakheit, 2004). The generally poor prognosis of the severe forms of poststroke language impairment coupled with the limited effectiveness of conventional speech and language therapy has stimulated the search for other treatments that may be used in conjunction with speech and language therapy, including the use of various drugs; for example, piracetam (Nootropil), amphetamines, and more recently donepezil (Aricept) (Bakheit, 2004). The justification for the use of drugs in the treatment of aphasia is based on evidence (Bakheit, 2004) that dextro-amphetamine (dexedrine) improves attention span and enhances learning and memory (Bakheit, 2004). Learning is an essential mechanism for the acquisition of new motor and cognitive skills, and hence, for recovery from aphasia. And preclinical and clinical data suggest that drug treatment may partially restore the metabolic function in the ischemic zone that surrounds the brain lesion and also has a neuro-protective effect following acute brain damage. Also animal studies have demonstrated the beneficial effects of this and other drugs on neural plasticity, though data on humans are still sparse (Bakheit, 2004). Certainly, a nonfluent aphasia case treated successfully with speech therapy and adjunctive mixed amphetamine salts was reported (Spiegel and Alexander, 2011). On the other hand, de Boissezon et al. (2007) highlight that pharmacotherapy of aphasia had been discussed for the last 20 years with first bromocriptine and amphetamine and then serotonergic, GABAergic, and cholinergic agents. These authors conclude that so far, proofs of efficiency were found indubitable for none of the studied molecules; however, some of them showed limited efficiency (piracetam and amphetamine). Moreover, drug therapies for aphasia were less efficient alone than when they were associated with speech therapy (de Boissezon et al., 2007); certainly, caution should be considered about d-amphetamine use (Yu, 2012).

Within the catecholamines, DA is the first catecholamine synthesized from DOPA. In turn, norepinephrine and epinephrine are derived from further metabolic modification of DA. McGaugh (2013) reminds us that although forgetting is the common fate of most of our experiences, much evidence indicates that emotional arousal enhances the storage of memories, thus serving to create, selectively, lasting memories of our more important experiences. The neurobiological systems mediating emotional arousal and memory are very closely linked. The adrenal stress hormones epinephrine (a catecholamine) and corticosterone released by emotional arousal regulate the consolidation of long-term memory (LTM). The amygdala plays a critical role in mediating these stress hormone influences (McGaugh, 2013). The release of norepinephrine in the amygdala and the activation of noradrenergic receptors are essential for stress-hormone-induced memory enhancement. The findings of both animal and human studies provide compelling evidence that stress-induced activation of the amygdala and its interactions with other brain regions involved in processing memory play a critical role in ensuring that emotionally significant experiences are well remembered (McGaugh, 2013).

Serotonin

Other neurotransmitter involved in memory is 5-HT. Indeed, drugs acting through 5-hydroxytryptamine (serotonin or 5-HT) systems modulate memory and its alterations, although the mechanisms involved are poorly understood. Even more in the context of the notion of memory deficits in psychiatry disorders (Millan et al., 2012). 5-HT still continues to generate interest as one of the most successful targets for therapeutic applications (e.g., depression, schizophrenia, anxiety, learning and memory disorders) (Bonsi et al., 2007; Lyseng-Williamson, 2013; Meltzer, 2013; Nordquist and Oreland, 2010). Indeed, growing evidence supports the notion that serotonergic is involved in memory, and this notion has gained wider acceptance and interest (Berger et al., 2009; Engelborghs et al., 2013; Rodríguez et al., 2012). Perhaps an important advantage of 5-HT consists that it has diverse pharmacological and genetic tools, neurotoxins, receptor agonist, and antagonists (e.g. Hoyer et al., 1994) and a well studied signaling and synaptic modulation (Bockaert et al., 2006; Lesch and Waider, 2012; Raymond et al., 2001; 2006). Certainly, in the last few years a growing number of papers had appeared that directly or indirectly implicate 5-HT systems in learning and memory in species ranging from humans to invertebrates (Rajasethupathy et al., 2012), and this trend continues in May 2013 (6119 total papers). An important insight provided by PubMed showing that in 1960 two papers were published while in 2012 354 publications appeared, with a peak (370 papers) in 2008. Notably, Monje et al. (2013) using a translational approach—from invertebrates to rodents—reported an evolutionary-conserved memory-related protein upregulated in implicit and explicit learning paradigms.

5-HT Systems and Neurobiological Markers Related to Memory Systems

The identification of $5-HT_1$ to $5-HT_7$ receptor families and its transporter in mammalian species and drugs selective for these sites (Fink and Göthert, 2007; Hoyer et al., 1994, 2002; Tohyama et al., 1998) have allowed to dissect their participation in learning and memory (Altman and Normile, 1988; Dougherty and Oristaglio, 2013; King et al., 2008; Meneses, 1999, 2001, 2007a, 2007b; Nonkes et al., 2013; Noristani et al., 2012; Perez-Garcia and Meneses, 2008b; Puig and Gulledge, 2011; Roberts and Hedlund, 2012; Rosse and Schaffhauser, 2010; Rutsalainen et al., 1998; Steckler and Sahgal, 1995; Terry et al., 2008; Tsuruoka et al., 2012; Upton et al., 2008; Vitalis et al., 2013; Wilson and Terry, 2009). Growing evidence indicates that 5-HT receptors and (5-HT transporter) SERT are involved in normal, pathophysiological, and therapeutic aspects of learning and memory (Bosh et al., 2013; Gacsályi et al., 2013; Meneses, 1999). Notably, serotonin modulation and neural plasticity have a long history (Fletcher, 1997; Lesch and Waider, 2012). 5-HT drugs may present promnesic and/or antiamnesic effects. For instance, the effects of 5-HT endogenous on memory formation were studied by using a 5-HT uptake facilitator (tianeptine), facilitating memory consolidation and selective $5-HT_{1-7}$ receptor antagonists reversed it (Meneses, 2002) interacting with other neurotransmission systems (Meneses, 2002; and also Briand et al., 2007; Tellez et al., 2012a). Notably, disorders such as AD and schizophrenia have an important component of dysfunctional memory; their etiology includes dysfunction of cholinergic, glutamatergic, and serotonergic systems (Berger et al, 2009; Engelborghs et al., 2013; Rodríguez et al., 2012), and certainly 5-HT has been also implicated in diseases with memory disorders, including depression, compulsive disorder, and posttraumatic stress disorder (Hermann et al., 2012; Jones and Moller, 2011; Millan et al., 2012; Vermetten and Lanius, 2012) (Table 2.1).

5-HT Neural Markers and Memory

A timely question is if the 5-HT system is important in the context of neural markers and memory. The case of $5-HT_6$ receptor (Marazziti et al., 2011, 2013; Ramírez, 2013; Reid et al., 2010) provides an excellent and timely case. As rats overexpressing dorsomedial striatum, but not dorsocentral striatum, $5-HT_6$ receptor showed impaired

Table 2.1 5-HT Systems and Neurobiological Markers Related to Memory Systems

Functions and Dysfunctions	5-HT marker	References
Memory formation, aging, AD, and amnesia	SERT, 5-HT$_{1A-1D}$, 5-HT$_{2A/2C}$, 5-HT$_4$, 5-HT$_6$, and 5-HT$_7$ receptors $\downarrow$	Chou et al. (2012), Eppinger et al. (2012), Meneses (1999, 2003), Meneses and Perez-Garcia (2007), Rodríguez et al. (2012), Xu et al. (2012)
Memory deficits, promnesic and antiamnesic drugs	Modify 5-HT receptors and SERT	Belcher et al. (2005), Eriksson et al. (2012), Garcia-Alloza et al. (2004), Huerta-Rivas et al. (2010), Jones and Moller (2011), Lorke et al. (2006), Marcos et al. (2006, 2008), Mathur and Lovinger (2012), Meneses et al. (2007), Meneses et al. (2011a), Perez-Garcia and Meneses (2006, 2008a, 2009), Tellez et al. (2010)
Serotonergic manipulations alter memory	5-HT$_{1A}$, 5-HT$_{2A/2C}$, 5-HT$_3$, 5-HT$_4$, 5-HT$_6$, and 5-HT$_7$ receptors	Hindi Attar et al. (2012), Bockaert et al. (2008), Meneses (1999, 2003), King et al. (2008), Ögren et al. (2008), Roth et al. (2004), Terry et al. (2008)
Agonists and/or antagonists seem to have promnesic and/or antiamnesic effects	5-HT$_{1A}$, 5-HT$_4$, 5-HT$_6$, and 5-HT$_7$ receptors	Bockaert et al. (2008), Elvander-Tottie et al. (2009), Ivachtchenko and Ivanenkov (2012), Meneses (1999, 2003), King et al. (2008), Ögren et al. (2008), Roth et al. (2004), Ruiz and Oranias (2010), Terry et al. (2008), van Praag (2004), Youn et al. (2009)

performance in a simple operant learning task (a striatum-dependent learning model), but not in the hippocampus-dependent water maze task (Mitchell et al., 2007). This impairment effect was appreciable at third instrumental testing session or the second extinction session on performance of previously acquired instrumental conditioning (Mitchell et al., 2007). In contrast, during memory consolidation of Pavlovian/instrumental autoshaping learning task, requiring dentate gyrus, hippocampal CA1, basolateral amygdaloid nucleus, and PFC (Pérez-García and Meneses, 2008a; Tellez et al., 2010); memory formation required a serotonergic tone (at least) via 5-HT$_6$ receptor, suppressing its expression. Nevertheless, under amnesic conditions 5-HT$_6$ receptor is completely suppressed or slightly reduced. In contrast, when the selective 5-HT$_6$ receptor SB-399885 improved memory or amnesia was reversed, the expression of 5-HT$_6$ receptor was increased or reestablished, respectively. Moreover, memory formation on the water maze (MWM) downregulated 5-HT$_6$ receptor protein and mRNA receptor expression, and the administration of the selective 5-HT_6 receptor antagonist SB-271046 induced an increase in pCREB-1 levels while CREB-2 levels were significantly reduced (Marcos et al., 2010). However, although SB-271046 was able to improve retention in the MWM, no further changes in pCREB-1 or CREB-2 levels were observed. The MWM alone significantly increased pERK1/2 levels and further increases were seen when treating with SB-271046 during the MWM. Ly et al. (2013) had recently noted that the 5-HT_6 antagonist SAM-531, failed to potentiate theta power, which is characteristic of many procognitive substances, indicating that 5-HT_6 receptors might not tonically modulate hippocampal oscillations and sleep−wake patterns. Together this evidence offers us the following picture, which allows illustrating the useful of 5-HT systems.

Brain Areas, Biochemical Pathways, Cognitive-Enhancing Effects of 5-HT Receptor Drugs

Marcos et al. (2010) conclude that their results suggest that, in the hippocampus, biochemical pathways associated with pERK1/2 expression, and not with the CREB family of transcription factors, seem to be related to the cognitive-enhancing properties of 5-HT_6 receptor antagonists. Additionally, repeated treatment with the 5-HT_6 antagonist RO4368554 (5.0 mg/kg) improved novel object recognition (NOR) and social discrimination without changing plasticity-associated proteins Ki-67 and PCNA (Mitchell et al., 2009), suggesting that diverse molecular mechanisms may be associated to promnesic or antiamnesic effects of 5-HT_6 receptor antagonists. On the other hand, the 5-HT_6 receptor antagonists procognitive effects may be due to an indirect regulation of cholinergic (Bourson et al., 1995), glutamatergic (King et al., 2008), and/or even serotonergic transmission (Lieben et al., 2005; Hirst et al., 2006), through GABA release in PFC, hippocampus and striatum; important brain areas for memory and associative learning (Meneses, 2003; see also Packard and Knowlton, 2002; Squire and Zola, 1996; Zola-Morgan and Squire, 1993). Then serotonin clearly exerts a tonic effect upon 5-HT_6 receptor as shown by 5-HT_6 receptor antagonists' procognitive or antiamnesic effect. Even, in mechanistic terms, an interesting possibility is that 5-HT_6 receptor activates the extracellular signal-regulated kinase1/2 (ERK2) via a Fyn-dependent pathway, a member of the Src family of nonreceptor protein tyrosine kinases (see Choi et al., 2007; Yun et al., 2007) for references. Notably, Fyn is involved in AD through the modulation of microtubule-associated tau and amyloid; Fyn deficiency might be used in the diagnosis, development, and treatment of the AD. Notably, in a model of AD, acetylcholinesterase inhibitors and 5-HT_6 receptor antagonists were used to improve the model's prediction of clinical outcomes (Roberts et al., 2012).

In addition, 5-HT_6 receptor agonists might be useful for their promemory and/or antiamnesic effects (Woods et al., 2012) acting on the cholinergic and/or glutamatergic neurons or alternatively; 5-HT_6 receptor antagonists and agonists might operate via modulation of distinct intracellular signaling pathways (de Foubert et al., 2007; Liu and Robichaud, 2009; Romero et al., 2006; Woods et al., 2012). As already noted (see above), density and 5-HT$_6$ levels were significantly decreased in a cohort of AD patients (Marcos et al., 2008). In the following lines other serotonergic neurobiological markers showing changes related to AD and/or aging.

5-HT Pathways, Receptors, and Transporter: Memory Functions and Dysfunctions

5-HT pathways project to almost all brain areas (Hoyer et al., 1994; Tohhyama et al., 1998), and diverse 5-HT mechanisms might be useful in the treatment of learning and memory dysfunctions. Notably, aging, AD, and amnesia are associated to decrements in 5-HT markers such as SERT and in the number of 5-HT$_{1A-1B}$, 5-HT$_{2A/2C}$, 5-HT$_4$, 5-HT$_6$, and 5-HT$_7$ receptors (for references see Meneses, 1999; Rodríguez et al., 2012; Xu et al., 2012). Also emerging evidence indicates that memory formation, amnesia, promnesic and amnesic drugs modify serotonergic markers, including 5-HT receptors, SERT, and serotonergic tone (Belcher et al., 2005; Callaghan et al., 2012; Da Silva Costa-Aze et al., 2012; Eriksson et al., 2012; Fournet et al., 2012; Frankland et al., 2013; Froestl et al., 2012, 2013; Garcia-Alloza et al., 2004; Goldenhuys and Van der Schyf, 2009; Gong et al., 2012; Haahr et al., 2012; Hermann et al., 2012; Huerta-Rivas et al., 2010; Jones and Moller, 2011; Lorke et al., 2006; Liy-Salmeron, 2008; Marin et al., 2012a; Marin et al., 2012b; Marcos et al., 2006, 2008; Marshall and O'Dell, 2012; Mathur and Lovinger, 2012; Meneses et al., 2007, 2011a; Na et al., 2012; Pérez-García et al., 2006; Pérez-García and Meneses, 2008a; Pérez-García and Meneses, 2009; Tellez et al., 2010, 2012). Certainly, in this context, 5-HT systems, protocols of training/testing, memory tasks and drugs (see e.g., Meneses, 2013) deserve attention.

Diverse techniques have been useful in the identification of serotonergic and/or other neurotransmitters (Ersche et al., 2011; Meyer, 2012) markers accompanying cognitive processes, including memory formation and memory disorders, ranging from autoradiography, RT-PCR, Western blot, etc. (for review see Meneses and Liy-Salmeron (2012); see also Martin and Sibson, 2008). As diverse serotonergic manipulations alter memory (Hindi Attar et al., 2012), certainly, 5-HT$_{1A}$, 5-HT$_{2A/2C}$, 5-HT$_3$, 5-HT$_4$, 5-HT$_6$, and 5-HT$_7$ receptors have attracted more scientific interest regarding memory (Bockaert et al., 2008, 2011; Boulougouris V and Robbins, 2010; Cammarota et al., 2008; Cowen and Sherwood, 2013; Meneses, 1999, 2003; King et al., 2008; Ögren et al., 2008; Roth et al., 2004; Terry et al., 2008; Volk et al., 2010; Williams et al., 2002). In addition, 5-HT$_{1A/1B}$, 5-HT$_{2A/2C}$, 5-HT$_3$, 5-HT$_4$, 5-HT$_6$, and 5-HT$_7$ receptors have in common that their agonists and/or antagonists seem to have promnesic and/or antiamnesic effects (Bockaert et al., 2008; Borg, 2008; Bombardi and Di Giovanni, 2013; Elvander-Tottie et al., 2009; Hajjo et al., 2012; Ivachtchenko and Ivanenkov, 2012; Ivachtchenko et al., 2012; Meneses, 1999, 2003; King et al., 2008; Ögren et al., 2008; Pennanen et al., 2013; Roth et al., 2004; Ruiz and Oranias, 2010; Sawyer et al., 2012; Terry et al., 2008; Thompson et al., 2013; Timotijević et al., 2012; van Praag, 2008; Youn et al., 2009; Yun and Rhim, 2011a, 2011b; Yun et al., 2010; Zhang et al., 2013). Certainly this apparent paradox remains for clarification.

In conclusion, the role of 5-HT systems on memory has become a major area of scientific interest, 5-HT$_6$ and 5-HT$_7$ receptor agonists and/or antagonists represent the more recent focus. This situation is not new regarding serotonergic receptors (e.g., 5-HT$_{1A}$ receptor). And the identification of reliable neural markers is fundamental for the understanding of memory mechanisms, its alterations, and potential treatment and cure (?). Considering a genuine need to treat cognitive deficits associated with many neuropsychiatric conditions as well as an increasingly aging population; including diverse neurotransmission systems cell surface receptors and intracellular drug targets.

Protocols of Training/Testing, Memory Tasks, and Drugs

Multiple behavioral tasks exist for studying memory. For instance, PubMed (June, 2, 2013) reveals 1195 papers, showing that in 1974 two related papers were published while 69 in 2012. Figure 2.1 depicts diverse behavioral tasks (Peele and

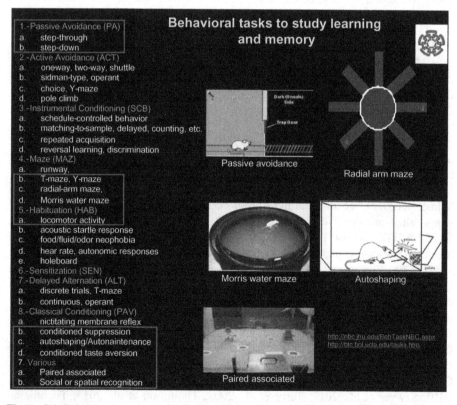

Figure 2.1 Behavioral memory models are shown, illustrating some. Modified after Vincent and Peele (1989).

Vincent, 1989; Lynch, 2004; Gupta et al., 2012; McGaugh, 1989; Myhrer, 2003; Sarter et al., 1992; Talpos et al., 2013; Tomi et al., 2012). Perhaps, among the more popular behavioral tasks are passive avoidance, the Morris water maze, radial maze, spontaneous alternation, and NOR. Figure 2.2 aims drawing attention to the cognitive demand, behavioral requirements, and brain areas involved in those tasks. Which should provide a rationale for using behavioral tasks resting on the notion that the hippocampus (together with rhinal and parahipocampal cortices) is paramount for memory formation. Certainly, a heuristic new approach might include other brain areas and their chronological participation.

Other important considerations should include type of memory, the dynamic hierarchy of neural markers, and brain areas involved in memory formation (Euston et al., 2012) versus no training, amnesia, antiamnesic effects, or forgetting (Tellez et al., 2012). Likewise, the species and the nature of behavioral task (e.g., appetitivly or aversively motivated), curves of behavioral acquisition (i.e., multitrial or two trials task) or patterns of behavioral responses (progressive vs. all or none response), cognitive demand (easy or difficult task), timing of drug administration (pretraining, posttraining, or pretest) and kind of drug (e.g., agonist or antagonist), protocols of training and testing together with neurobiological markers accompanying mnemonic processes deserve attention. All these factors have produced similar results, and they are responsible for some inconsistencies.

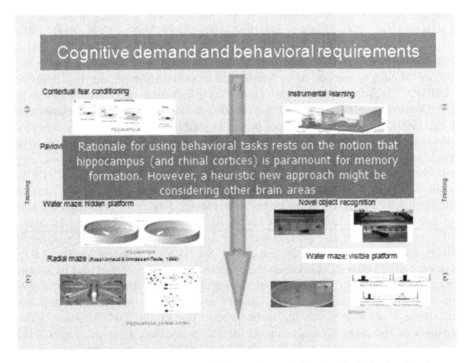

Figure 2.2 Some behavioral memory models are shown, indicating the behavioral and cognitive (up to down) increasing demand involved.

Revision of them might provide further insights about the consistent and variability in results among laboratories, using different or similar memory tasks, animals, etc. Certainly, it is a complex and multifocal issue; hence, herein we are addressing a few aspects using some examples, which should provide an analytic framework offering some clues. For instance, when comparing 5-HT$_6$ receptor drugs, memory tasks, and protocols of training, similar and some differential outcomes are observed between, for example, appetitive associative Pavlovian/instrumental autoshaping versus aversively motivated water Maze (spatial memory) and inhibitory avoidance (Meneses et al., 2011b). For instance, the development of potent and selective 5-HT$_6$ receptor antagonists has been crucial in the clarification of the role of this receptor on memory. Even the evidence that 5-HT$_6$ receptor agonists can improve memory is providing new insights. In this context, clarifications about whether 5-HT receptor drugs are acting as inverse agonists/antagonists might be very crucial. Importantly, determination if drugs, age and/or memory alone alter 5-HT$_6$ receptor expression and/or signal cascades is very important. Actually, memory tasks modify the expression of neurobiological markers (e.g., serotonergic receptors, SERT, signaling) (Marcos et al., 2010; Tellez et al., 2012a, 2012b). Likewise, even, different instruments for measuring memory also are relevant (see e.g., Gonzalez et al., 2013).

Moreover, the analysis of behavioral specificity, drug administration, and memory changes sheds new light on the finding that while the effects of the dual 5-HT$_{1A/7}$ receptor agonist 8-OHDPAT on learning and memory were unclear (Calcagno et al., 2006; Carli et al., 1992, 1995, 1997, 1999a, 199b, 2000, 2001, 2006; Eriksson et al., 2012; Haider et al., 2012; Lladó-Pelfort et al., 2012; Millan et al., 2004; Perez-Garcia and Meneses, 2007; Ögren et al., 2008; Schechter et al., 2005). It should be noticed nevertheless, that a consistent result was low doses of 8-OHDPAT facilitated memory in several tasks, including autoshaped response in mice (Vanover and Barrett, 1998) and rats (Pérez-García et al., 2006; Meneses and Hong, 1999), while its higher doses produced memory deficits in diverse memory tasks (Meneses and Perez-Garcia, 2007), including the Morris water maze, passive avoidance, and autoshaping. Using the same memory task or different memory tasks had produced diverse data (autoshaping see e.g., Stahlman et al., 2010; Tomie et al., 2012). Notably, the implementation of memory task or the inherent instruments for measuring memory becomes crucial. Regarding autoshaping learning task contrasting data had been found, for example, between Meneses et al., 2011a and Lindner et al. (2003). Indeed the latter authors did not find significant effects with the 5-HT$_6$ receptor antagonist Ro 04-6790 (Lindner et al., 2003). Important differences exist between these works (e.g., the lever used and its illumination; Meneses et al., 2009). It should be noted that new instruments for measuring behavior in autoshaping task (Bussey et al., 2012) are being useful for addressing issues like variability intersubjects (Cook et al., 2004; Meneses, 2003; Meneses et al., 2009; Gonzalez et al., 2013). In addition, Gravius et al. (2011) reported data contrasting with Schreiber et al. (2007); in their data the former reported moderate efficacy in some models for AD but do not support the therapeutic potential of 5-HT$_6$ antagonists for schizophrenia.

Loci, Mechanisms of Action, and Memory Tasks

Centrally and/or systemically administered drugs (Meneses et al., 2009; Menard and Treit, 1999) and genetic manipulations offer excellent opportunities in the identification and localization of mechanisms or loci. Indeed, administration of 8-OHDAP into the dorsal raphe had no effect on water maze but compensated the deficit on spatial learning caused by impaired cholinergic or glutamatergic hippocampal transmission (Carli et al., 2001) and it enhanced operant conditional discrimination (Ward et al., 1999). On operant autoshaping task mice lacking 5-HT$_{1A}$ or 5-HT$_{1B}$ receptors learned faster (Pattij, 2002), and even the latter mice showed water maze-enhanced platform acquisition and probe trial performance. Likewise, selective 5-HT$_{1B}$ receptor antagonists facilitated memory in the water maze (Skelton et al., 2008), as well as did 5-HT-moduline a peptide endogenous (Moret et al., 2003) in autoshaped memory (Hong et al., 1999). In parallel to the finding that 5-HT$_{2B/2C}$ receptors antagonist had no effect on autoshaped memory (Meneses, 2003), 5-HT$_{2C}$ receptor knockout mice showed normal hidden platform and preference for the target quadrant during a probe trial (for references see Meneses, 2003) or with repeated training in a 10-day win-shift radial arm maze task, 5-HT$_{2C}$ knock-out and wild type (WT) mice showed similar decreases of the number of working memory and reference memory errors (Hill et al., 2011).

Other important examples are the case of 5-HT$_3$ or 5-HT$_4$ receptor. Thus, antagonists for 5-HT$_3$ receptor (Hodges et al., 1996) or agonists for 5-HT$_4$ receptor attenuated water maze or autoshaping deficits in scopolamine-treated and/or forebrain-lesioned rats or atropine-induced memory deficits, respectively (Hong and Meneses, 1996; for references see also Meneses and Hong, 1997). In passive avoidance tasks (for references see Meneses, 2003; Meneses and Perez-Garcia, 2007) acquisition, memory consolidation and retrieval were impaired by 5-HT$_{1A}$ or 5-HT$_{1B}$ (Skelton et al., 2008) receptor stimulation, while blockade of 5-HT$_{1A}$ or 5-HT$_{2A/2B/2C}$ receptor had no effect or improved performance, respectively. In addition, 5-HT$_3$ antagonists or 5-HT$_4$ agonists (systemic or centrally administered) improved passive avoidance learning and/or prevented scopolamine- or hypoxic-induced amnesia. Exploration of possible brain structures revealed that, for example, 5-HT posttraining injections into the dorsal and ventral aspects of striatum selectively produced strong amnesia on passive avoidance, illustrating that striatal 5-HT activity is involved in memory functions and neurochemical heterogeneity within the striatum regarding memory consolidation (Prado-Alcalá et al., 2003).

Memory Tasks and Signaling

These are six key steps in the molecular biological delineation of short-term memory (STM) and its conversion to LTM for both implicit (procedural) and explicit (declarative) memory, namely cAMP (cyclic adenosine monophosphate), PKA (protein kinase A), CRE (the cAMP response element), CREB-1 (CRE-binding protein or cAMP response element binding protein-1; a transcriptional activator that promotes memory storage), CRE-binding protein (CREB-2, a repressor gene of memory), and the cytoplasmic polyadenylation element binding protein (CPEB) (Kandel, 2012; see also Dudanova and Klein, 2013). Moreover, Izquierdo et al. (2006; see also Vianna et al., 2000) have demonstrated (see also Izquierdo and McGaugh, 2000; McGaugh, 2013; McGaugh and Izquierdo, 2000) that cAMP/protein kinase C (PKC) changes are both necessary for the initiation and continuity of LTM consolidation on passive avoidance task, which are modulated by diverse neurotransmission systems and transduction pathways, including 5-HT$_{1A}$ synapses in the CA1 hippocampal area, entorhinal and posterior parietal cortex (Izquierdo and McGaugh, 2000). Memory consolidation in passive avoidance was associated to increased cAMP production (Izquierdo et al., 2006). In this regard, the dual 5-HT$_{1A/7}$ agonist 8-OHDPAT-facilitatory effect on autoshaped memory (LTM 24 and 48 h) was accompanied by cAMP increased cortical and hippocampal cAMP production (Manuel-Apolinar and Meneses, 2004; Meneses, 2002). The selective antagonists WAY100635 (5-HT$_{1A}$) or DR4004 (5-HT$_7$ receptor; Kikuchi et al., 2003) alone had no effect on memory. In the interaction experiments, the 8-OH-DPAT−WAY100635 or 8-OH-DPAT−DR4004 combinations, did not modify or enhanced 8-OH-DPAT-induced increased cAMP, respectively (Manuel-Apolinar and Meneses, 2004). It should be noted that, Izquierdo et al. (2006) found augmented cAMP production (immediately and 90 min later) and memory formation in passive avoidance task.

A further analysis (Perez-Garcia and Meneses, 2008; Perez-Garcia and Meneses, 2008a) of STM (1.5 h) and LTM (24 and 48 h) protocol and considering the data of untrained and treated animals as the basal values of cAMP production relative to those of trained and treated animals showed that memory formation (saline group) *per se* increased cAMP production relative to untrained animals. Memory formation plus 8-OHDPAT elicited lower values of cAMP in PFC; but in the untrained group, cAMP was increased (Perez-Garcia and Meneses, 2008; Perez-Garcia and Menses, 2008a). In consequence, cAMP was increased or decreased by the stimulation (8-OHDPAT) of 5-HT$_{1A/7}$ receptor if memory was improving but the opposite occurred in the lack of memory (Perez-Garcia and Meneses, 2008; Perez-Garcia and Menses, 2008a). As the 5-HT$_7$ receptor agonist AS19 improved memory and elicited higher (raphe nuclei and PFC) cAMP production in both trained and untrained animals, its intrahippocampal administration produced even significant higher cAMP scores in all three brain areas (Perez-Garcia and Meneses, 2008; Perez-Garcia and Menses, 2008a). Effects of 8-OHDPAT and AS19 were partially or completely blockade by WAY 100635 or SB269970, respectively; thus, providing further support to the notion that 5-HT$_{1A}$ and 5-HT$_7$ receptors were involved in these effects (Perez-Garcia and Menses, 2008a, 2008b). Notably, different protocols of training/testing and memory tasks seem to be associated to cAMP production and memory formation. Importantly, it is unclear the 5-HT$_{1A}$ and 5-HT$_7$ receptors contribution, of both, or other neurotransmitters and signaling. These data allow illustrating the importance and complexity of cAMP production and drug administration in the signaling in mammalian memory formation. The analysis of the interaction among brain areas, neurotransmitters systems (Briand et al., 2007) shows that the expression of diverse transporters (e.g., GABA, glutamate, serotonin, etc.) is modulated by memory formation, amnesia, forgetting (Tellez et al., 2012a, 2012b).

Brain Areas, Neurotransmitters Systems, Drugs: Cognitive and Behavioral Demand of Memory Tasks and Protocols of Training: a Final Consideration

Memory formation requires hippocampus in behavioral tasks such as water maze (Kandel, 2001), passive avoidance (Izquierdo et al., 1999), and Pavlovian/instrumental autoshaping (Meneses, 2003; Meneses et al., 2009; Tellez et al., 2010). In contrast, overtrained animals engage more striatum and less hippocampus on autoshaping (Meneses, 2003; Perez-Garcia and Meneses, 2009; Tellez et al., 2010) and passive avoidance (Izquierdo et al., 2006). It should be observed that while explicit or declarative memory has been related to hippocampus, implicit or nondeclarative memory has been related to striatum (Adamantidis and de Lecea, 2009; Kandel, 2001). Nevertheless, it should be considered that multiple memory mechanisms can work in tandem to support performance on an implicit memory task, and even additional contribution of explicit memory can be observed in neurologically healthy individuals (Koenig et al., 2009) or during acquisition of an otherwise implicit learning task (Meneses et al., 2011). This implicates that during the early stage (i.e., learning) of implicit memory, it does not depend on striatum; nevertheless, as training and time go through implicit memory eventually depends on striatum (Izquierdo et al., 2006). Thus it seems to be appropriate to revise the notion that, regardless of the time of training/testing explicit and implicit memories, are mediated by hippocampus and striatum, respectively. This is an important consideration in as much as Buchhave et al. (2012) had highlighted that approximately 90% of patients with MCI and pathologic CSF biomarker levels at baseline develop AD within 9−10 years. Levels of Aβ42 are already fully decreased at least 5−10 years before conversion to AD dementia, whereas T-tau and P-tau seem to be later markers. These results provide direct support in humans for the hypothesis that altered Aβ metabolism precedes tau-related pathology and neuronal degeneration (Buchhave et al., 2012; Wiener et al., 2012).

Signaling and Memory

Molecular signaling is key field of scientific investigation (see Chapter 8). Also, particularly, as the study of the molecular bases of memory has been become a field of the major scientific interest (e.g., Davis and Squire, 1984; Kandel, 2001; Means, 2008), hence, in the following lines some aspects are mentioned. Thus, initial studies of the molecular switch from STM to LTM in Aplysia and Drosophila focused on regulators like CREB-1 that promote memory storage (Bekinschtein et al., 2010; Kandel, 2012). However, subsequent studies in Aplysia and in the fly revealed the surprising finding that the switch to long-term synaptic change and the growth of new synaptic connections is also constrained by memory suppressor genes (Costa-Mattioli and Sonenberg, 2008; Kandel, 2012; Mayford et al., 2012). One important memory suppressor gene that constrains the growth of new synaptic connections is CREB-2, which when overexpressed blocks long-term synaptic facilitation in Aplysia. When CREB-2 is removed, a single exposure to serotonin, which normally produces an increase in synaptic strength lasting only minutes, will increase synaptic strength for days and induce the robust growth of new synaptic connections (Kandel, 2012). The CREB-mediated response to extracellular stimuli can be modulated by a number of kinases (PKA, CaMKII, CaMKIV, RSK2, MAPK, and PKC) and phosphatases (PP1 and calcineurin). The CREB regulatory unit may therefore serve to integrate signals from various signal transduction pathways. This ability to integrate signaling, as well as to mediate activation or repression, may explain why CREB is so central to memory storage (Kandel, 2012). These transcriptional repressors and activators can interact with each other both physically and functionally. It is likely that the transition is a complex process involving temporally distinct phases of gene activation, repression, and regulation of signal transduction (Kandel, 2012). In addition to protein kinases, synaptic protein phosphatases also play a key role in regulating the initiation of long-term synaptic changes. Various protein phosphatases, such as PP1 and calcineurin, counteract the local activity of PKA acting as inhibitory constraints of memory formation. For example, calcineurin can act as a memory suppressor for sensitization in the Aplysia. An equilibrium between kinase and phosphatase activities, at a given synapse gates the synaptic signals that reach the nucleus and thus, can regulate both memory storage and retrieval (Kandel, 2012).

In short, according with Kandel (2012), the analysis of the contributions to synaptic plasticity and memory of cAMP, PKA, CRE, CREB-1, CREB-2, and CPEB has recruited the efforts of many laboratories all over the world. These are six key steps in the molecular biological delineation of STM and its conversion to LTM both implicit (procedural) and explicit (declarative) memory (Kandel, 2012). Notably, Monje et al. (2013) reported that studies of synaptic plasticity using the marine mollusk *Aplysia californica* as model system have been successfully used to identify proteins involved in learning and memory. The importance of molecular elements regulated by the learning-related neurotransmitter serotonin in Aplysia can then be explored in rodent models and finally tested for their relevance for human physiology and pathology (Monje et al., 2013). For instance, Flotillin-1, a member of the flotillin/reggie family of scaffolding proteins, has been previously found to be involved in neuritic branching and synapse formation in hippocampal neurons *in vitro*. Importantly, elevated levels of Flotillin-1 in hippocampal tissue of mice trained in the Morris water maze confirmed the relevance of Flotillin-1 for memory-related processes in a mammalian system. Hence, translational approach—from invertebrates to rodents—led to the identification of Flotillin-1 as evolutionary-conserved memory-related protein (Monje et al., 2013).

Memories and Molecular Traces

Importantly, as already mentioned some 5-HT drugs may present promnesic and/or antiamnesic effects (King et al., 2008; Terry et al., 2008). Perhaps an important advantage of 5-HT consists that it has diverse pharmacological and genetic tools, neurotoxins, receptor agonist and antagonists (see e.g., Meneses and Liy-Salmeron, 2012) and a well studied signaling and synaptic modulation (Bockaert et al., 2006; Lesch and Waider, 2012; Turner et al., 2007). Hence, in the following lines recent evidence about memories and molecular traces is revised.

It is often stated that STM is consolidated in a protein-synthesis-dependent manner into LTM ((e.g., Davis and Squire, 1984), Sossin, 2008). Alternatively, memories might consist of distinct molecular traces that last for different periods of time; these traces can be graded by their "volatility," traces encoded by activation of protein kinases are more volatile than traces encoded by morphological changes at preexisting synapses (Sossin, 2008; Squire and Davies, 1981). The least volatile ("static") traces are due to the generation and stabilization of new synapses; importantly, whereas at the cellular level these traces are generated independently of each other, they might be linked at the network level where volatile memory traces are required to set up a cellular network that is in turn required to induce the static memory trace (Sossin, 2008).

Sweatt (2009) see also Telese et al., 2013 describes recent discoveries demonstrating that experience can drive the production of epigenetic marks in the adult nervous system and that the experience-dependent regulation of epigenetic (mechanisms that allow a heritable change in gene expression in the absence of DNA mutation; Rahn et al., 2013; Roopra et al., 2012) molecular mechanisms in the mature central nervous system participates in the control of gene transcription underlying the formation of LTMs. In the mammalian experimental systems investigated thus far, epigenetic mechanisms have been linked to associative fear conditioning, extinction of learned fear, and hippocampus-dependent spatial memory formation (Sweatt, 2009). Intriguingly, in one experimental system epigenetic marks at the level of chromatin structure (histone acetylation; e.g., Gräef and Tsai, 2013) have been linked to the recovery of memories that had seemed to be "lost" (i.e., not available for recollection) (Sweatt, 2009). Environmental enrichment has long been known to have positive effects on memory capacity, and recent studies have suggested that these effects are at least partly due to the recruitment of epigenetic mechanisms (Sweatt, 2009). Finally, an uncoupling of signal transduction pathways from the regulation of epigenetic mechanisms in the nucleus has been implicated in the closure of developmental critical periods. According with Sweatt (2009), taken together, these eclectic findings suggest a new perspective on experience-dependent dynamic regulation of epigenetic mechanisms in the adult nervous system and their relevance to biological psychiatry. Notably, Mansour et al (2013) have proposed the link chronic pain and the role of learning and brain plasticity

According with Guan et al. (2002) although much is known about short-term integration, little is known about how neurons sum opposing signals for long-term synaptic plasticity and memory storage. In Aplysia, Guan et al. (2002) found that when a sensory neuron simultaneously receives inputs from the facilitatory transmitter 5-HT at one set of synapses and the inhibitory transmitter FMRFamide at another, long-term facilitation is blocked and synapse-specific long-term depression dominates. Chromatin immunoprecipitation assays show that 5-HT induces the downstream gene C/EBP by activating CREB-1, which recruits CBP for histone acetylation, whereas FMRFa leads to CREB-1 displacement by CREB-2 and recruitment of HDAC5 to deacetylate histones. When the two transmitters are applied together, facilitation is blocked because CREB-2 and HDAC5 displace CREB1-CBP, thereby deacetylating histones (Guan et al., 2002). Rahn et al. (2013) characterize epigenetic mechanisms as critical for the gene expression profile necessary to induce and maintain long-lasting neuronal plasticity and behavior; broadly defined epigenetic mechanisms are a set of processes and modifications, influencing gene function without alteration of the primary DNA sequence. Canonical epigenetic mechanisms include histone posttranslational modifications (PTMs) and DNA methylation, although recent research has also identified a number of other processes involved in epigenetic regulation including noncoding RNAs, prions, chromosome position effects, and Polycomb repressors (Rahn et al., 2013). Notably, Jarome and Lubin (2013) highlight that histone lysine methylation is a well-established transcriptional mechanism for the regulation of gene expression changes in eukaryotic cells and is now believed to function in neurons of the central nervous system to mediate the process of memory formation and behavior. In mature neurons, methylation of histone proteins can serve to both activate and repress gene transcription. This is in stark contrast to other epigenetic modifications, including histone acetylation and DNA methylation, which have largely been associated with one transcriptional state in the brain. Jarome and Lubin (2013) discuss the evidence for histone methylation mechanisms in the coordination of complex cognitive processes such as LTM formation and storage. In addition, the current literature highlighting the role of histone methylation in intellectual disability, addiction, schizophrenia, autism, depression, and neurodegeneration (Jarome and Lubin, 2013); discussing histone methylation within the context of other epigenetic modifications and the potential advantages of exploring this newly identified mechanism of cognition, emphasizing the possibility that this molecular process may provide an alternative locus for intervention in long-term psychopathologies that cannot be clearly linked to genes or environment alone (Jarome and Lubin, 2013).

As already mentioned, LTM formation requires transcription and protein synthesis (Peixoto and Abel, 2013; see Chapter 7). Over the past few decades, a great amount of knowledge has been gained regarding the molecular players that regulate the transcriptional program linked to memory consolidation (Peixoto and Abel, 2013). Epigenetic mechanisms have been shown to be essential for the regulation of neuronal gene expression, and histone acetylation has been one of the most studied and best characterized. Peixoto and Abel (2013) summarize lines of evidence that have shown the relevance of histone acetylation in memory in both physiological and pathological conditions. Great advances have been made in identifying the writers and erasers of histone acetylation marks during learning; however, the identities of the upstream regulators and downstream targets that mediate the effect of changes in histone acetylation during memory consolidation remain restricted to a handful of molecules (Peixoto and Abel, 2013). Peixoto and Abel (2013) outline a general model by which corepressors and coactivators regulate histone acetylation during memory storage and discuss how the recent advances in high-throughput sequencing have the potential to radically change our understanding of how epigenetic control operates in the brain (Peixoto and Abel, 2013).

Indeed, long-lasting memories require specific gene expression programs that are, in part, orchestrated by epigenetic mechanisms (Gräff and Tsai, 2013). Of the epigenetic modifications identified in cognitive processes, as mentioned above, histone acetylation has spurred considerable interest. Whereas increments in histone acetylation have consistently been shown to favor learning and memory, a lack thereof has been causally implicated in cognitive impairments in neurodevelopmental disorders, neurodegeneration, and aging (Gräff and Tsai, 2013). As histone acetylation and cognitive functions can be pharmacologically restored by histone deacetylase inhibitors, this epigenetic modification might constitute a molecular memory aid on the chromatin and, by extension, a new template for therapeutic interventions against cognitive frailty (Gräff and Tsai, 2013).

Moreover, Thellier and Lüttge (2013) consider that all memory functions have molecular bases, namely in signal reception and transduction, and in storage and recall of information. As at all levels of organization, living organisms have some kind of memory (Thellier and Lüttge, 2013). An interesting example, in plants one may distinguish two types (Thellier and Lüttge, 2013); there are linear pathways from reception of signals and propagation of effectors to a type of memory that may be described by terms such as learning, habituation, or priming. There is a storage and recall memory based on a complex network of elements with a high degree of integration and feedback. The most important elements envisaged are calcium waves, epigenetic modifications of DNA and histones, and regulation of timing via a biological clock. Thellier and Lüttge (2013) feature two sorts of memory and show how those can be distinguished, proposing a schematic model of plant memory, which is derived as emergent from integration of the various modules. Possessing the two forms of memory supports the fitness of plants in response to environmental stimuli and stress (Thellier and Lüttge, 2013). This evidence offers new perspectives concerning investigation memory. Notably, Mansour et al. (2013) have proposed the link chronic pain and the role of learning and brain plasticity.

On other hand, the ongoing quest for memory enhancement is an important investigation area, considering that the global population is increasingly aging (Stern and Alberini, 2012). The extraordinary progress that has been made in the past few decades elucidating the underlying mechanisms of how LTMs are formed has provided insight into how memories might also be enhanced (Stern and Alberini, 2012). Capitalizing on this knowledge, it has been postulated that targeting many of the same mechanisms, including CREB activation, AMPA/NMDAR trafficking, neuromodulation (e.g., via DA, adrenaline, cortisol, or acetylcholine), and metabolic processes (e.g., via glucose and insulin), may all lead to the enhancement of memory (Stern and Alberini, 2012). These and other mechanisms and/or approaches have been tested via genetic or pharmacological methods in animal models, and several have been investigated in humans as well (Stern and Alberini, 2012). In addition, a number of behavioral methods, including exercise and reconsolidation, may also serve to strengthen and enhance memories. By utilizing this information and continuing to investigate these promising avenues, memory enhancement may indeed be achieved in the future (Stern and Alberini, 2012). Very importantly, Terry et al. (2011) postulated that the rise in elderly populations is also resulting in an increase in individuals with related (potentially treatable) conditions such as "MCI" which is characterized by a less severe (but abnormal) level of cognitive impairment and a high risk for developing dementia. Even in the absence of a diagnosable disorder of cognition (e.g., AD and MCI), the perception of increased forgetfulness and declining mental function is a clear source of apprehension in the elderly (Terry et al., 2011). This is a valid concern given that even a modest impairment of cognitive function is likely to be associated with significant disability in a rapidly evolving, technology-based society. Unfortunately, the currently available therapies designed to improve cognition (i.e., for AD and other forms of dementia) are limited by modest efficacy and adverse side effects, and their effects on cognitive function are not sustained over time (Terry et al., 2011). Accordingly, it is incumbent on the scientific community to develop safer and more effective therapies that improve and/ or sustain cognitive function in the elderly allowing them to remain mentally active and productive for as long as possible. As diagnostic criteria for memory disorders evolve, the demand for procognitive therapeutic agents is likely to surpass AD and dementia to include MCI and potentially even less severe forms of memory decline (Terry et al., 2011). Terry et al. (2011) (see also Wallace et al., 2011) provide an overview of the contemporary therapeutic targets and preclinical pharmacologic approaches (with representative drug examples) designed to enhance memory function, including cholinergic agents (nicotinic and muscarinic), phosphodiesterase inhibitors, serotonergic drugs, histamine H_3 receptor antagonists, multiple drug targets and multifunctional compounds and additional multifunctional compounds.

Importantly, Stern and Alberini (2012) (see also Lynch et al., 2011) mention that in addition to pharmacological approaches, there are a number of behavioral manipulations that have been found to be effective in promoting memory enhancement. First, repetition has long been known to enhance memory performance, and it has been consistently shown that repeated training trials and/or learning events are associated with better memory. This method is one that is employed during everyday learning and is also known to have benefits for cognitive disorders such as dementia. Similarly, memory can be enhanced by targeting retrieval-induced reconsolidation, which occurs through repeated retrieval sessions. When a memory is retrieved, it can again return to a labile state and can undergo reconsolidation. It has been suggested that a function of reconsolidation is to increase memory strength (Stern and Alberini, 2012). Notably memory changes over time, and in order to attain a memory enhancing effect, retrievals must occur within a relatively short time span after training. Memory strengthening is a function of the age/stage of the memory and that likewise, memory storage is dynamic and changes over time (Stern and Alberini, 2012). However, there is currently little knowledge about the mechanisms by which these changes occur, or whether the behavioral effect based on reconsolidation would be useful in aging and/or AD. Interestingly, the reconsolidation process does not promote memory enhancement by behavioral repetitions only. Memory enhancement can be in fact promoted via pharmacological manipulations given in concert with reconsolidation. Thus, injections of nicotine, β-adrenergic receptor agonists, PKA activators, phosphodiesterase type 5 inhibitors, angiotensin, and IGF-II all enhance memory when injected after retrieval (Stern and Alberini, 2012). This suggests that memories can be enhanced even after they are consolidated and indicates potential new directions for developing treatments for cognitive disorders that targets in addition to deficits of encoding/consolidation and deficits of already-formed memories (Stern and Alberini, 2012). Another behavioral approach that has been recently found to be effective in enhancing memories is physical exercise. Numerous studies over the past two decades in rodents have shown conclusively that exercise can improve memory in a number of tasks, including tasks such as spatial water maze, fear-based and the nonaversive NOR task. This effect may be more pronounced in aged rats and has numerous possible underlying mechanisms, including increase of BDNF, neurogenesis, IGF-I, glucocorticoids, and CAMKII activation, suggesting once again that mechanisms involved in memory consolidation may be the best targets to achieve memory enhancement or prevent memory decay (Stern and Alberini, 2012). Though it is generally accepted that exercise is beneficial in regards to general health, extended life span and aging, the extent and type of exercise needed in humans to attain a benefit specifically for memory enhancement is not yet mechanistically well understood, and excessive forms of exercise may actually be deleterious for declarative, hippocampal-dependent memories by causing extreme stress. Behavioral methods for enhancing memory are an exciting avenue of research and may be extremely useful in clinical practice with more basic knowledge of how best to implement their practice, as well as whether they can be even further augmented through pharmacological means, as with reconsolidation-mediated enhancement (Stern and Alberini, 2012).

Importantly, activity-dependent changes in synaptic strength are considered mechanisms underlying learning and memory (Yamada and Nabeshima, 2003). Brain-derived neurotrophic factor (BDNF) plays an important role in activity-dependent synaptic plasticity. Indeed, memory acquisition and consolidation are associated with an increase in BDNF mRNA expression and the activation of its receptor TrkB, and genetic as well as pharmacologic deprivation of BDNF or TrkB impairs learning and memory (Yamada and Shimada, 2003). In a positively motivated radial arm maze test, activation of the TrkB/phosphatidylinositol-3 kinase (PI3-K) signaling pathway in the hippocampus is associated with consolidation of spatial memory through an activation of translational processes. In a negatively motivated passive avoidance test, mitogen-activated protein kinase (MAPK) is activated during acquisition of fear memory (Yamada and Shimada, 2003). Also, the interaction between BDNF/TrkB signaling and NMDARs for spatial memory. A Src-family tyrosine kinase, Fyn plays a role in this interaction by linking TrkB with NR2B. These findings suggest that BDNF/TrkB signaling in the hippocampus plays a crucial role in learning and memory (Yamada and Shimada, 2003).

Also, the Ca(2+)/Calmodulin(CaM)-dependent protein kinase II (CaMKII) (Coultrap and Bayer, 2012) is activated by Ca(2+)/CaM, but becomes partially autonomous (Ca(2+)-independent) upon autophosphorylation at T286. This hallmark feature of CaMKII regulation provides a form of molecular memory and is indeed important in LTP of excitatory synapse strength and memory formation. According with Coultrap and Bayer (2012), however, emerging evidence supports a direct role in information processing, while storage of synaptic information may instead be mediated by regulated interaction of CaMKII with the NMDAR complex. These and other CaMKII regulation mechanisms are discussed by Coultrap and Bayer (2012) in the context of the kinase structure and their impact on postsynaptic functions. Recent findings also implicate CaMKII in LTD, as well as functional roles at inhibitory synapses, lending renewed emphasis on better understanding the spatiotemporal control of CaMKII regulation (Coultrap and Bayer, 2012).

Finally, several of the neurotransmitter systems and their associated signaling had been related to memory process (see e.g., glutamate, Kandel, 2001, 2012; GABA, Paille et al., 2013; cholinergic, Hernandez and Dineley, 2012; serotonergic, King et al., 2008; Marcos et al., 2010; Millan et al., 2012). For instance, progress in understanding the complex biology of DA receptor-related signal transduction mechanisms has revealed that, in addition to their primary action on cAMP-mediated signaling, DA receptors (as other neurotransmission systems) can act through diverse signaling mechanisms that involve alternative G-protein coupling or through G-protein-independent mechanisms via interactions with ion channels or proteins that are characteristically implicated in receptor desensitization, such as β-arrestins (Beaulieu and Gainetdinov, 2011). One of the future directions in managing DA-related pathologic conditions may involve a transition from the approaches that directly affect receptor function to a precise targeting of postreceptor intracellular signaling modalities either directly or through ligand-biased signaling pharmacology (Beaulieu and Gainetdinov, 2011).

Acknowledgments

I thank Sofia Meneses-Goytia for revised language and Roberto Gonzalez for his expert assistance. This work was partially supported by CONACYT grant 80060.

References

Adamantidis, A., de Lecea, L., 2009. A role for melanin-concentrating hormone in learning and memory. Peptides. 30, 2066–2070.

Altman, H.J., Normile, H.J., 1988. What is the nature of the role of the serotonergic nervous system in learning and memory: prospects for development of an effective treatment strategy for senile dementia. Neurobiol. Aging. 9, 627−638.

Anggono, V., Huganir, R.L., 2012. Regulation of AMPA receptor trafficking and synaptic plasticity. Curr. Opin. Neurobiol. 22 (3), 461−469.

Atack, J.R., 2011. GABAA receptor subtype-selective modulators. II. α5-selective inverse agonists for cognition enhancement. Curr. Top Med. Chem. 11 (9), 1203−1214.

Bakheit, A.M., 2004. Drug treatment of poststroke aphasia. Expert Rev. Neurother. 4 (2), 211−217.

Beaulieu, J.M., Gainetdinov, R.R., 2011. The physiology, signaling, and pharmacology of dopamine receptors. Pharmacol. Rev. 63 (1), 182−217.

Bekinschtein, P., Katche, C., Slipczuk, L., Gonzalez, C., Dorman, G., Cammarota, M., et al., 2010. Persistence of long-term memory storage: new insights into its molecular signatures in the hippocampus and related structures. Neurotox. Res. 18 (3−4), 377−385.

Belcher, A.M., O'Dell, S.J., Marshall, J.F., 2005. Impaired object recognition memory following methamphetamine, but not p-chloroamphetamine- or d-amphetamine-induced neurotoxicity. Neuropsychopharmacology. 30 (11), 2026−2034.

Bentley, P., Driver, J., Dolan, R.J., 2011. Cholinergic modulation of cognition: insights from human pharmacological functional neuroimaging. Prog. Neurobiol. 94 (4), 360−388.

Berger, M., Gray, J.A., Roth, B.L., 2009. The expanded biology of serotonin. Annu. Rev. Med. 60, 355−366.

Bockaert, J., Claeysen, S., Bécamel, C., Dumuis, A., Marin, P., 2006. Neuronal 5-HT metabotropic receptors: fine-tuning of their structure, signaling, and roles in synaptic modulation. Cell Tissue Res. 326, 553−572.

Bockaert, J., Claeysen, S., Compan, V., Dumuis, A., 2008. 5-HT$_4$ Receptors: history, molecular pharmacology and brain functions. Neuropharmacol. 55, 922−931.

Bockaert, J., Claeysen, S., Compan, V., Dumuis, A., 2011. 5-HT$_4$ Receptors, a place in the sun: act two. Curr. Opin. Pharmacol. 11 (1), 87−93.

Bombardi, C., Di Giovanni, G., 2013. Functional anatomy of 5-HT2A receptors in the amygdala and hippocampal complex: relevance to memory functions. Exp. Brain Res. 230 (4), 427−439.

Bonsi, P., Cuomo, D., Ding, J., Sciamanna, G., Ulrich, S., Tscherter, A., et al., 2007. Endogenous serotonin excites striatal cholinergic interneurons via the activation of 5-HT 2C, 5-HT6, and 5-HT7 serotonin receptors: implications for extrapyramidal side effects of serotonin reuptake inhibitors. Neuropsychopharmacology 32 (8), 1840−1854.

Borg, J., 2008. Molecular imaging of the 5-HT$_{1A}$ receptor in relation to human cognition. Behav. Brain Res. 195 (1), 103−111.

Bosch, O.G., Wagner, M., Jessen, F., Kühn, K.U., Joe, A., Seifritz, E., et al., 2013. Verbal memory deficits are correlated with prefrontal hypometabolism in (18)FDG PET of recreational MDMA users. PLoS One. 8 (4), e61234.

Boulougouris, V., Robbins, T.W., 2010. Enhancement of spatial reversal learning by 5-HT$_{2C}$ receptor antagonism is neuroanatomically specific. J. Neurosci. 30 (3), 930−938.

Bourson, A., Borroni, E., Austin, R.H., Monsma Jr, F.J., Sleight, A.J., 1995. Determination of the role of the 5-ht6 receptor in the rat brain: a study using antisense oligonucleotides. J. Pharmacol Exp. Ther. 274 (1), 173−180.

Briand, L.A., Gritton, H., Howe, W.M., Young, D.A., Sarter, M., 2007. Modulators in concert for cognition: modulator interactions in the prefrontal cortex. Prog. Neurobiol. 83 (2), 69−91.

Buchhave, P., Minthon, L., Zetterberg, H., Wallin, A.K., Blennow, K., Hansson, O., 2012. Cerebrospinal fluid levels of β-amyloid 1-42, but not of tau, are fully changed already 5 to 10 years before the onset of Alzheimer dementia. Arch. Gen. Psychiatry. 69 (1), 98–106.

Bussey, T.J., Holmes, A., Lyon, L., Mar, A.C., McAllister, K.A., Nithianantharajah, J., et al., 2012. New translational assays for preclinical modelling of cognition in schizophrenia: the touchscreen testing method for mice and rats. Neuropharmacology 62 (3), 1191–1203.

Calcagno, E., Carli, M., Invernizzi, R.W., 2006. The 5-HT$_{1A}$ receptor agonist 8-OH-DPAT prevents prefrontocortical glutamate and serotonin release in response to blockade of cortical NMDA receptors. J. Neurochem. 96 (3), 853–860.

Callaghan, C.K., Hok, V., Della-Chiesa, A., Virley, D.J., Upton, N., O'Mara, S.M., 2012. Age-related declines in delayed non-match-to-sample performance (DNMS) are reversed by the novel 5HT$_6$ receptor antagonist SB742457. Neuropharmacology 63 (5), 890–897.

Cammarota, M., Bevilaqua, L.R., Medina, J.H., Izquierdo, I., 2008. ERK1/2 and CaMKII-mediated events in memory formation: is 5HT regulation involved? Behav. Brain Res. 195, 120–128.

Carli, M., Samanin, R., 1992. 8-Hydroxy-2- (di-n-propylamino) tetralin impairs spatial learning in a water maze: role of postsynaptic 5-HT1A. Br. J. Pharmacol. 105, 720–726.

Carli, M., Luschi, R., Garofalo, P., Samanin, R., 1995. 8-OH-DPAT impairs spatial but not visual learning in a water maze by stimulating 5-HT$_{1A}$ receptors in the hippocampus. Behav. Brain Res. 67, 67–74.

Carli, M., Bonalumi, P., Samanin, R., 1997. WAY 100635, a 5-HT$_{1A}$ receptor antagonist, prevents the impairment of spatial learning caused by intrahippocampal administration of scopolamine or 7-chloro-kynurenic acid. Brain Res. 774, 167–174.

Carli, M., Bonalumi, P., Samanin, R., 1998. Stimulation of 5-HT$_{1A}$ receptors in the dorsal raphe reverses the impairment caused by intrahippocampal scopolamine in rats. Eur. J. Neurosci. 10, 221–230.

Carli, M., Silva, S., Balducci, C., Samanin, R., 1999a. WAY 100635, a 5-HT$_{1A}$ receptor antagonist, prevents the impairment of spatial learning caused by blockade of hippocampal NMDA receptors. Neuropharmacology. 38, 1165–1173.

Carli, M., Balducci, C., Millan, M.J., Bonalumi, P., Samanin, R., 1999b. S 15535, a benzodioxopiperazine acting as presynaptic agonist and postsynaptic 5-HT$_{1A}$ receptor antagonist, prevents the impairment of spatial learning caused by intrahippocampal scopolamine. Br. J. Pharmacol. 128, 1207–1214.

Carli, M., Samanin, R., 2000. The 5-HT$_{1A}$ receptor agonist 8-OH-DPAT reduces rats' accuracy of attentional performance and enhances impulsive responding in a five-choice serial reaction time task: role of presynaptic 5-HT$_{1A}$ receptors. Psychopharmacology (Berl). 149, 259–268.

Carli, M., Balducci, C., Samanin, R., 2001. Stimulation of 5-HT$_{1A}$ receptors in the dorsal raphe ameliorates the impairment of spatial learning caused by intrahippocampal 7-chloro-kynurenic acid in naive and pretrained rats. Psychopharmacology (Berl). 158 (1), 39–47.

Carli, M., Baviera, M., Invernizzi, R.W., Balducci, C., 2006. Dissociable contribution of 5-HT$_{1A}$ and 5-HT$_{2A}$ receptors in the medial prefrontal cortex to different aspects of executive control such as impulsivity and compulsive perseveration in rats. Neuropsychopharmacol. 31, 757–767.

Choi, Y.H., Kang, H., Lee, W.K., Kim, T., Rhim, H., Yu, Y.G., 2007. An inhibitory compound against the interaction between G alpha(s) and the third intracellular loop region of serotonin receptor subtype 6 (5-HT6) disrupts the signaling pathway of 5-HT6. Exp. Mol. Med. 39 (3), 335−342.

Collingridge, G.L., Volianskis, A., Bannister, N., France, G., Hanna, L., Mercier, M., et al., 2013. The NMDA receptor as a target for cognitive enhancement. Neuropharmacology 64, 13−26.

Cook, R.G., Geller, A.I., Zhang, G.R., Gowda, R., 2004. Touchscreen-enhanced visual learning in rats. Behav. Res. Methods Instrum. Comput. 36 (1), 101−106.

Cooper, J.R., Bloom, F.E., Roth, R.H., 2003. The Biochemical Basis of Neuropharmacology. Oxford University Press, New York, NY.

Costa-Mattioli, M., Sonenberg, N., 2008. Translational control of gene expression: a molecular switch for memory storage. Prog. Brain Res. 169, 81−95.

Coultrap, S.J., Bayer, K.U., 2012. CaMKII regulation in information processing and storage. Trends Neurosci. 35 (10), 607−618.

Cowen, P., Sherwood, A.C., 2013. The role of serotonin in cognitive function: evidence from recent studies and implications for understanding depression. J. Psychopharmacol. 27 (7), 575−578.

Cummings, J., Gould, H., Zhong, K., 2012. Advances in designs for Alzheimer's disease clinical trials. Am. J. Neurodegener. Dis. 1 (3), 205−216.

Czubak, A., Nowakowska, E., Burda, K., Kus, K., Metelska, J., 2010. Cognitive effects of GABAergic antiepileptic drugs. Arzneimittelforschung. 60 (1), 1−11.

Da Silva Costa-Aze, V., Quiedeville, A., Boulouard, M., Dauphin, F., 2012. 5-HT6 Receptor blockade differentially affects scopolamine-induced deficits of working memory, recognition memory and aversive learning in mice. Psychopharmacology (Berl.). 222 (1), 99−115.

Dalley, J.W., Lääne, K., Theobald, D.E., Armstrong, H.C., Corlett, P.R., Chudasama, Y., et al., 2005. Time-limited modulation of appetitive Pavlovian memory by D1 and NMDA receptors in the nucleus accumbens. Proc. Natl. Acad. Sci. USA. 102 (17), 6189−6194.

Davis, H.P., Squire, L.R., 1984. Protein synthesis and memory: a review. Psychol. Bull. 96 (3), 518−559.

de Boissezon, X., Peran, P., de Boysson, C, Démonet, J.F., 2007. Pharmacotherapy of aphasia: myth or reality? Brain Lang. 102 (1), 114−125.

de Foubert, G., O'Neill, M.J., Zetterström, T.S., 2007. Acute onset by 5-HT(6)-receptor activation on rat brain brain-derived neurotrophic factor and activity-regulated cytoskeletal-associated protein mRNA expression. Neuroscience 147 (3), 778−785.

Decker, M.W., McGaugh, J.L., 2004. The role of interactions between the cholinergic system and other neuromodulatory systems in learning and memory. Synapse 7 (2), 151−168.

Dougherty, J.P., Oristaglio, J., 2013. Chronic treatment with the serotonin 2A/2C receptor antagonist SR 46349B enhances the retention and efficiency of rule-guided behavior in mice. Neurobiol. Learn. Memory 103, 50−63.

Dudanova, I., Klein, R., 2013. Integration of guidance cues: parallel signaling and crosstalk. Trends Pharmacol. Sci. 36 (5), 295−304.

Eichenbaum, H., 2013. What H.M. taught us. J. Cogn. Neurosci. 25 (1), 14−21.

Elvander-Tottie, E., Eriksson, T.M., Sandin, J., Ogren, S.O., 2009. 5-HT$_{1A}$ and NMDA receptors interact in the rat medial septum and modulate hippocampal-dependent spatial learning. Hippocampus 19 (12), 1187−1198.

Engelborghs, S., Sleegers, K., Van der Mussele, S., Le Bastard, N., Brouwers, N., Van Broeckhoven, C., et al., 2013. Brain-Specific tryptophan hydroxylase, TPH2, and 5-HTTLPR are associated with frontal lobe symptoms in Alzheimer's disease. J. Alzheimers Dis. 35 (1), 67–73.

Eppinger, B., Hämmerer, D., Li, S.C., 2012. Neuromodulation of reward-based learning and decision making in human aging. Ann. NY Acad. Sci. 1235, 1–17.

Eriksson, T.M., Holst, S., Stan, T.L., Hager, T., Sjögren, B., Ogren, S.O., et al., 2012. 5-HT$_{1A}$ and 5-HT$_7$ Receptor crosstalk in the regulation of emotional memory: Implications for effects of selective serotonin reuptake inhibitors. Neuropharmacology 63 (6), 1150–1160.

Ersche, K.D., Roiser, J.P., Lucas, M., Domenici, E., Robbins, T.W., Bullmore, E.T., 2011. Peripheral biomarkers of cognitive response to dopamine receptor agonist treatment. Psychopharmacology (Berl.). 214 (4), 779–789.

Euston, D.R., Gruber, A.J., McNaughton, B.L., 2012. The role of medial prefrontal cortex in memory and decision making. Neuron 76 (6), 1057–1070.

Farrant, M., 2001. Amino acids: inhibitory. In: Webster, R.A. (Ed.), Neurotransmitters, Drugs and Brain Function. John Wiley & Sons, Chichester, UK.

Fletcher, L., 1997. Memories are made of this: the genetic basis of memory. Mol. Med. Today 3 (10), 429–434.

Fink, K.B., Göthert, M., 2007. 5-HT Receptor regulation of neurotransmitter release. Pharmacol. Rev. 59 (4), 360–417.

Fournet, V., de Lavilléon, G., Schweitzer, A., Giros, B., Andrieux, A., Martres, M.P., 2012. Both chronic treatments by epothilone D and fluoxetine increase the short-term memory and differentially alter the mood-status of STOP/MAP6 KO mice. J. Neurochem. 123 (6), 982–996.

Frankland, P.W., Köhler, S., Josselyn, S.A., 2013. Hippocampal neurogenesis and forgetting. Trends Neurosci. 36 (9), 497–503.

Froestl, W., Muhs, A., Pfeifer, A., 2012. Cognitive enhancers (nootropics). Part 1: drugs interacting with receptors. J. Alzheimers Dis. 32 (4), 793–887.

Froestl, W., Pfeifer, A., Muhs, A., 2013. Cognitive enhancers (nootropics). Part 3: drugs interacting with targets other than receptors or enzymes. Disease-modifying drugs. J. Alzheimers Dis. 34 (1), 1–114.

Gacsályi, I., Nagy, K., Pallagi, K., Lévay, G., Hársing Jr, L.G., Móricz, K., et al., 2013. Egis-11150: a candidate antipsychotic compound with procognitive efficacy in rodents. Neuropharmacology 64, 254–263.

Garcia-Alloza, M., Hirst, W.D., Chen, C.P., Lasheras, B., Francis, P.T., Ramírez, M.J., 2004. Differential involvement of 5-HT$_{1B/1D}$ and 5-HT$_6$ receptors in cognitive and non-cognitive symptoms in Alzheimer's disease. Neuropsychopharmacology 29, 410–416.

Geldenhuys, W.J., Van der Schyf, C.J., 2009. The serotonin 5-HT$_6$ receptor: a viable drug target for treating cognitive deficits in Alzheimer's disease. Expert Rev. Neurother. 9, 1073–1085.

Gonda, X., 2012. Basic pharmacology of NMDA receptors. Curr. Pharm. Des. 18 (12), 1558–1567.

Gong, P., Zheng, Z., Chi, W., Lei, X., Wu, X., Chen, D., et al., 2012. An association study of the genetic polymorphisms in 13 neural plasticity-related genes with semantic and episodic memories. J. Mol. Neurosci. 46 (2), 352–361.

Graef, S., Schönknecht, P., Sabri, O., Hegerl, U., 2011. Cholinergic receptor subtypes and their role in cognition, emotion, and vigilance control: an overview of preclinical and clinical findings. Psychopharmacology (Berl.). 215 (2), 205−229, <10.1007/s00213-010-2153-8>. Epub 2011 Jan 8.

Gräff, J., Tsai, L.H., 2013. Histone acetylation: molecular mnemonics on the chromatin. Nat. Rev. Neurosci. 14 (2), 97−111.

Gravius, A., Laszy, J., Pietraszek, M., Sághy, K., Nagel, J., Chambon, C., et al., 2011. Effects of 5-HT6 antagonists, Ro-4368554 and SB-258585, in tests used for the detection of cognitive enhancement and antipsychotic-like activity. Behav. Pharmacol. 22 (2), 122−135.

Guan, Z., Giustetto, M., Lomvardas, S., Kim, J.H., Miniaci, M.C., Schwartz, J.H., et al., 2002. Integration of long-term-memory-related synaptic plasticity involves bidirectional regulation of gene expression and chromatin structure. Cell. 111 (4), 483−493.

Gupta, A., Hemraj, Jalhan, S., Jindal, A., Upmanyu, N., 2012. Various animal models to check learning and memory—a review. Int. J. Pharm. Pharm. Sci. 4 (Suppl. 3), 91−95.

Haahr, M.E., Fisher, P., Holst, K., Madsen, K., Jensen, C.G., Marner, L., et al., 2012. The 5-HT$_4$ receptor levels in hippocampus correlates inversely with memory test performance in humans. Hum. Brain Mapp. 34 (11), 3066−3074.

Haas, C., 2012. Strategies, development, and pitfalls of therapeutic options for Alzheimer's disease. J. Alzheimers Dis. 28 (2), 241−281.

Haider, S., Khaliq, S., Tabassum, S., Haleem, D.J., 2012. Role of somatodendritic and postsynaptic 5-HT$_{1A}$ receptors on learning and memory functions in rats, http://www.ncbi.nlm.nih.gov/pubmed/22814880. Neurochem. Res. 37 (10), 2161−2166.

Hajjo, R., Setola, V., Roth, B.L., Tropsha, A., 2012. Chemocentric informatics approach to drug discovery: identification and experimental validation of selective estrogen receptor modulators as ligands of 5-hydroxytryptamine-6 receptors and as potential cognition enhancers. J. Med. Chem. 55 (12), 5704−5719.

Hermann, A., Küpper, Y., Schmitz, A., Walter, B., Vaitl, D., Hennig, J., et al., 2012. Functional gene polymorphisms in the serotonin system and traumatic life events modulate the neural basis of fear acquisition and extinction. PLoS One. 7 (9), e44352.

Hernandez, C.M., Dineley, K.T., 2012. α7 Nicotinic acetylcholine receptors in Alzheimer's disease: neuroprotective, neurotrophic or both? Curr. Drug Targets. 13 (5), 613−622.

Hill, R.A., Murray, S.S., Halley, P.G., Binder, M.D., Martin, S.J., van den Buuse, M., 2011. Brain-derived neurotrophic factor expression is increased in the hippocampus of 5-HT$_{2C}$ receptor knockout mice. Hippocampus 21 (4), 434−445.

Hindi Attar, C., Finckh, B., Büchel, C., 2012. The influence of serotonin on fear learning. PLoS One. 7 (8), c42397.

Hirst, W.D., Stean, T.O., Rogers, D.C., Sunter, D., Pugh, P., Moss, S.F., et al., 2006. SB-399885 is a potent, selective 5-HT$_6$ receptor antagonist with cognitive enhancing properties in aged rat water maze and novel object recognition models. Eur. J. Pharmacol. 553 (1−3), 109−119.

Hong, E, Orozco, G, Meneses, A, Fillion, G., 1999. Effect of 5-HT-moduline, an endogenous peptide, in associative learning. Proc. West. Pharmacol. Soc. 42, 37−38.

Hong, E., Meneses, A., 1996. Systemic injection of p-chloroamphetamine eliminates the effect of the 5-HT$_3$ compounds on learning. Pharmacol. Biochem. Behav. 53 (4), 765−769.

Hodges, H., Sowinski, P., Turner, J.J., Fletcher, A., 1996. Comparison of the effects of the 5-HT$_3$ receptor antagonists WAY-100579 and ondansetron on spatial learning in the water maze in rats with excitotoxic lesions of the forebrain cholinergic projection system. Psychopharmacol (Berl). 125 (2), 146−161.

Hoyer, D., Clarke, D.E., Fozard, J.R., Hartig, P.R., Martin, G.R., Mylecharane, E.J., et al., 1994. International union of pharmacology classification of receptors for 5-hydroxytryptamine (Serotonin). Pharmacol. Rev. 46 (2), 157−203.

Hoyer, D., Hannon, J.P., Martin, G.R., 2002. Molecular, pharmacological and functional diversity of 5-HT receptors. Pharmacol. Biochem. Behav. 7, 533−554.

Huerta-Rivas, A., Perez-Garcia, G., Gonzalez, C., Meneses, A., 2010. Time-course of 5-HT$_6$ receptor mRNA expression during memory consolidation and amnesia. Neurobiol. Learn. Mem. 93 (1), 99−110.

Ivachtchenko, A.V., Ivanenkov, Y.A., 2012. 5HT6 receptor antagonists: a patent update. Part 1. Sulfonyl derivatives. Expert Opin. Ther. Pat. 22 (8), 917−964.

Ivachtchenko, A.V., Ivanenkov, Y.A., Skorenko, A.V., 2012. 5-HT6 Receptor modulators: a patent update. Part 2. Diversity in heterocyclic scaffolds. Expert Opin. Ther. Pat. 22 (10), 1123−1168.

Izquierdo, I., Cunha, C., Medina, J.H., 1990. Endogenous benzodiazepine modulation of memory processes. Neurosci. Biobehav. Rev. 14 (4), 419−424.

Izquierdo, I., Medina, J.H., Vianna, M.R., Izquierdo, L.A., Barros, D.M., 1999. Separate mechanisms for short- and long-term memory. Behav. Brain Res. 103 (1), 1−11.

Izquierdo, I., Bevilaqua, L.R., Rossato, J.I., Bonini, J.S., Medina, J.H., Cammarota, M., 2006. Different molecular cascades in different sites of the brain control memory consolidation. Trends Neurosci. 29 (9), 496−505.

Izquierdo, I., McGaugh, J.L., 2000. Behavioural pharmacology and its contribution to the molecular basis of memory consolidation. Behav. Pharmacol. 11 (7−8), 517−534.

Jane, D.E., Lodge, D., Collingridge, G.L., 2009. Kainate receptors: pharmacology, function and therapeutic potential. Neuropharmacology 56 (1), 90−113.

Jarome, T.J., Lubin, F.D., 2013. Histone lysine methylation: critical regulator of memory and behavior. Rev. Neurosci. 27, 1−13.

Jones, T., Moller, M.D., 2011. Implications of hypothalamic-pituitary-adrenal axis functioning in posttraumatic stress disorder. J. Am. Psychiatry Nurses Assoc. 17 (6), 393−403.

Kandel, E.R., 2001. The molecular biology of memory storage: a dialogue between genes and synapses. Science 294, 1030−1038.

Kandel, E.R., 2012. The molecular biology of memory: cAMP, PKA, CRE, CREB-1, CREB-2, and CPEB. Mol. Brain. 5, 14.

Kemp, A., Manahan-Vaughan, D., 2007. Hippocampal long-term depression: master or minion in declarative memory processes? Trends Neurosci. 30 (3), 111−118.

Kikuchi, C., Suzuki, H., Hiranuma, T., Koyama, M., 2003. New tetrahydrobenzindoles as potent and selective 5-HT$_7$ antagonists with increased In vitro metabolic stability. Bioorg. Med. Chem. Lett. 13 (1), 61−64.

King, M.V., Marsden, C.A., Fone, K.C., 2008. A role for the 5-HT$_{1A}$, 5-HT$_4$ and 5-HT$_6$ receptors in learning and memory. Trends Pharmacol. Sci. 29, 482−492.

Klinkenberg, I., Blokland, A., 2010. The validity of scopolamine as a pharmacological model for cognitive impairment: a review of animal behavioral studies. Neurosci. Biobehav. Rev. 34 (8), 1307−1350.

Koenig, P., Smith, E.E., Troiani, V., Anderson, C., Moore, P., Grossman, M., 2009. Medial temporal lobe involvement in an implicit memory task: evidence of collaborating implicit and explicit memory systems from FMRI and Alzheimer's disease. Cereb Cortex. 18, 2831−2843.

Koh, M.T., 2012. Pharmacological treatment strategies targeting cognitive impairment associated with aging. Nihon Yakurigaku Zasshi. 139 (4), 157−159 (Article in Japanese).

Lanctôt, K.L., Herrmann, N., Black, S.E., Ryan, M., Rothenburg, L.S., Liu, B.A., et al., 2008. Apathy associated with Alzheimer disease: use of dextroamphetamine challenge. Am. J. Geriatr. Psychiatry. 16 (7), 551−557.

Lesch, K.P., Waider, J., 2012. Serotonin in the modulation of neural plasticity and networks: implications for neurodevelopmental disorders. Neuron 76 (1), 175−191.

Lieben, C.K., Blokland, A., Sik, A., Sung, E., van Nieuwenhuizen, P., Schreiber, R., 2005. The selective 5-HT$_6$ receptor antagonist Ro4368554 restores memory performance in cholinergic and serotonergic models of memory deficiency in the rat. Neuropsychopharmacology 30 (12), 2169−2179.

Lindner, M.D., Hodges Jr, D.B., Hogan, J.B., Orie, A.F., Corsa, J.A., Barten, D.M., et al., 2003. An assessment of the effects of serotonin 6 (5-HT6) receptor antagonists in rodent models of learning. J. Pharmacol. Exp. Ther. 307 (2), 682−691.

Liu, G.L., Robichaud, A.J., 2009. 5-HT$_6$ Antagonists as potential treatment for cognitive dysfunction. Drug Dev. Res. 70, 145−168.

Liy-Salmeron, G., Meneses, A., 2008. Role of 5-HT$_{1-7}$ receptors in short- and long-term memory for an autoshaping task: intrahippocampal manipulations. Brain Res. 1147, 140−147.

Lladó-Pelfort, L., Santana, N., Ghisi, V., Artigas, F., Celada, P., 2012. 5-HT$_{1A}$ receptor agonists enhance pyramidal cell firing in prefrontal cortex through a preferential action on GABA interneurons. Cereb. Cortex 22 (7), 1487−1497.

Lorke, D.E., Lu, G., Cho, E., Yew, D.T., 2006. Serotonin 5-HT$_{2A}$ and 5-HT$_6$ receptors in the prefrontal cortex of Alzheimer and normal aging patients. BMC Neurosci. 7, 36.

Ly, S., Pishdari, B., Lok, L.L., Hajos, M., Kocsis, B., 2013. Activation of 5-HT$_6$ receptors modulates sleep−wake activity and hippocampal theta oscillation. ACS Chem. 4, 191−199.

Lynch, G., 2006. Glutamate-based therapeutic approaches: ampakines. Curr. Opin. Pharmacol. 6 (1), 82−88.

Lynch, G., Palmer, L.C., Gall, C.M., 2011. The likelihood of cognitive enhancement. Pharmacol. Biochem. Behav. 99 (2), 116−129.

Lynch, M.A., 2004. Long term potentiation. Physiol. Rev. 84, 87−136.

Lyseng-Williamson, K., 2013. Clozapine: a guide to its use in patients with schizophrenia who are unresponsive to or intolerant of other antipsychotic agents. Drug Ther. Perspect. 29, 161−165.

Maffei, A., 2011. The many forms and functions of long term plasticity at GABAergic synapses. Neural Plast. 254724, 9 pages.

Mansour, A.R., Farmer, M.A., Baliki, M.N., Apkarian, A.V., 2013. Chronic pain: the role of learning and brain plasticity. Restor Neurol. Neurosci. [Epub ahead of print] PMID: 23603439.

Manuel-Apolinar, L., Meneses, A., 2004. 8-OH-DPAT facilitated memory consolidation and increased hippocampal and cortical cAMP production. Behav. Brain Res. 148 (1−2), 179−184.

Marazziti, D., Baroni, S., Catena Dell'Osso, M., Bordi, F, Borsini, F., 2011. Serotonin receptors of type 6 (5-HT6): what can we expect from them? Curr. Med. Chem. 18 (18), 2783−2790.

Marazziti, D., Baroni, S., Borsini, F., Picchetti, M., Vatteroni, E., Falaschi, V., et al., 2013. Serotonin receptors of type 6 (5-HT6): from neuroscience to clinical pharmacology. Curr. Med. Chem. 20 (3), 371−377.

Marcos, B., Gil-Bea, F.J., Hirst, W.D., García-Alloza, M., Ramírez, M.J., 2006. Lack of localization of 5-HT$_6$ receptors on cholinergic neurons: implication of multiple neurotransmitter systems in 5-HT$_6$ receptor-mediated acetylcholine release. Eur. J. Neurosci. 24, 299−1306.

Marcos, B., García-Alloza, M., Gil-Bea, F.J., Chuang, T.T., Francis, P.T., Chen, C.P., et al., 2008. Involvement of an altered 5-HT$_6$ receptor function in behavioral symptoms of Alzheimer's disease. J. Alzheimers Dis. 14, 43−50.

Marcos, B., Cabero, M., Solas, M., Aisa, B., Ramirez, M.J., 2010. Signaling pathways associated with 5-HT$_6$ receptors: relevance for cognitive effects. Int. J. Neuropsychopharmacol. 9, 1−10.

Marin P., Meffre J., Chaumont-Dubel S., La Cour C.L., Loiseau F., Watson D.J.G., et al., 2012a. 5-HT6 Receptors disrupt cognition by recruiting mTOR: relevance to schizophrenia. Serotonin Club meeting, Abstract p. 58. July 10−12, Montpellier, France.

Marin, P., Becamel, C., Dumuis, A., Bockaert, J., 2012b. 5-HT Receptor-associated protein networks: new targets for drug discovery in psychiatric disorders? Curr. Drug Targets. 13 (1), 28−52.

Marshall, J.F., O'Dell, S.J., 2012. Methamphetamine influences on brain and behavior: unsafe at any speed? Trends Neurosci. 35 (9), 536−545.

Martin, C., Sibson, N.R., 2008. Pharmacological MRI in animal models: a useful tool for 5-HT research? Neuropharmacology 55, 1038−1047.

Martyn, A.C., De Jaeger, X., Magalhães, A.C., Kesarwani, R., Gonçalves, D.F., Raulic, S., et al., 2012. Elimination of the vesicular acetylcholine transporter in the forebrain causes hyperactivity and deficits in spatial memory and long-term potentiation. Proc. Natl. Acad. Sci. USA 109 (43), 17651−17656.

Mathur, B.N., Lovinger, D.M., 2012. Serotonergic action on dorsal striatal function. Parkinsonism Relat. Disord. 18 (Suppl. 1), S129−S131.

Mayford, M., Siegelbaum, S.A., Kandel, E.R., 2012. Synapses and memory storage. Cold. Spring. Harb. Perspect. Biol. 4 (6), 1−18.

McGaugh, J.L., 1973. Drug facilitation of learning and memory. Annu. Rev. Pharmacol. 13, 229−241.

McGaugh, J.L., 1989. Dissociating learning and performance: drug and hormone enhancement of memory storage. Brain Res. Bull. 23 (4−5), 339−345.

McGaugh, J.L., 2006. Make mild moments memorable: add a little arousal. Trends Cogn. Sci. 10, 345−347.

McGaugh, J.L., 2013. Making lasting memories: remembering the significant. Proc. Natl. Acad. Sci. USA.110 (Suppl 2), 10402−10407.

McGaugh, J.L., Izquierdo, I., 2000. The contribution of pharmacology to research on the mechanisms of memory formation. Trends Pharmacol. Sci. 21 (6), 208−210.

Means, A.R., 2008. The year in basic science: calmodulin kinase cascades. Mol. Endocrinol. 22 (12), 2759−2765.

Menard, J., Treit, D., 1999. Effects of centrally administered anxiolytic compounds in animal models of anxiety. Neurosci. Biobehav. Rev. 23 (4), 591−613.

Meltzer, H.Y., 2013. Update on typical and atypical antipsychotic drugs. Annu. Rev. Med. 64, 393−406.

Meneses, A., 1999. 5-HT System and cognition. Neurosci. Biobehav. Rev. 23, 1111−1125.

Meneses, A., 2001. Effects of the 5-HT$_6$ receptor antagonist Ro 04-6790 on learning consolidation. Behav. Brain Res. 118 (1), 107−110.

Meneses, A., 2002. Tianeptine: 5-HT uptake sites and 5-HT$_{(1-7)}$ receptors modulate memory formation in an autoshaping Pavlovian/instrumental task. Neurosci. Biobehav. Rev. 26 (3), 309−319.

Meneses, A., 2003. Pharmacological analysis of an associative learning task: 5-HT$_1$ to 5-HT$_7$ receptor subtypes function on a Pavlovian/instrumental autoshaped memory. Learn. Mem. 10, 363−372.

Meneses, A., 2007a. Stimulation of 5-HT$_{1A}$, 5-HT$_{1B}$, 5-HT$_{2A/2A}$, 5-HT$_3$ and 5-HT$_4$ receptors or 5-HT uptake inhibition: short- and long-term memory. Behav. Brain Res. 184, 81−90.

Meneses, A., 2007b. Do serotonin$_{1-7}$ receptors modulate short and long-term memory? Neurobiol. Learn. Mem. 87, 561−572.

Meneses, A, Hong, E., 1999. 5-HT$_{1A}$ receptors modulate the consolidation of learning in normal and cognitively impaired rats. Neurobiol. Learn. Mem. 71 (2), 207−218.

Meneses, A., Hong, E., 1997. A pharmacological analysis of serotonergic receptors: effects of their activation of blockade in learning. Prog. Neuropsychopharmacol. Biol. Psychiatry. 21 (2), 273−296.

Meneses, A., 2013. 5-HT systems: emergent targets for memory formation and memory alterations. Rev. Neurosci (in press).

Meneses, A., Liy-Salmeron, G., 2012. Serotonin and emotion, learning and memory. Rev. Neurosci. 23, 443−453.

Meneses, A., Perez-Garcia, G., 2007. 5-HT$_{1A}$ Receptors and memory. Neurosci. Biobehav. Rev. 31 (5), 705−727.

Meneses, A., Manuel-Apolinar, L., Castillo, C., Castillo, E., 2007. Memory consolidation and amnesia modify 5-HT$_6$ receptors expression in rat brain: an autoradiographic study. Behav. Brain Res. 178, 53−61.

Meneses, A., Perez-Garcia, G., Liy-Salmeron, G., Ponce-Lopez, T., Tellez, R., Flores-Galvez, D., 2009. Associative learning, memory and serotonin: a neurobiological and pharmacological analysis. In: Rocha Arrieta, L.L., Granados-Soto, V. (Eds.), Models of Neuropharmacology. Transworld Research Network, Trivandrum, India, 978-81-7895-383-0 pp. 169−182.

Meneses, A., Pérez-García, G., Ponce-Lopez, T., Castillo, C., 2011a. 5-HT$_6$ Receptor memory and amnesia: behavioral pharmacology—learning and memory processes. Int. Rev. Neurobiol. 96, 27−47.

Meneses, A., Perez-Garcia, G., Ponce-Lopez, T., Tellez, R., Castillo, C., 2011b. Serotonin transporter and memory. Neuropharmacology 61 (3), 355−363.

Meyer, J.H., 2012. Neuroimaging markers of cellular function in major depressive disorder: implications for therapeutics, personalized medicine, and prevention. Clin. Pharmacol. Ther. 91 (2), 201−214.

Millan, M.J., Gobert, A., Roux, S., Porsolt, R., Meneses, A., Carli, M., et al., 2004. The serotonin1A receptor partial agonist S15535 [4-(benzodioxan-5-yl)1-(indan-2-yl) piperazine] enhances cholinergic transmission and cognitive function in rodents: a combined neurochemical and behavioral analysis. J. Pharmacol. Exp. Ther. 311 (1), 190−203.

Millan, M.J., Agid, Y., Brüne, M., Bullmore, E.T., Carter, C.S., Clayton, N.S., et al., 2012. Cognitive dysfunction in psychiatric disorders: characteristics, causes and the quest for improved therapy. Nat. Rev. Drug Discov. 11 (2), 141−168.

Mitchell, E.S., Sexton, T., Neumaier, J.F., 2007. Increased expression of 5-HT$_6$ receptors in the rat dorsomedial striatum impairs instrumental learning. Neuropsychopharmacology 32, 1520−1530.

Mitchell, E.S., McDevitt, R.A., Neumaier, J.F., 2009. Adaptations in 5-HT receptor expression and function: implications for treatment of cognitive impairment in aging. J. Neurosci. Res. 87 (12), 2803−2811.

Monje, F.J., Divisch, I., Demit, M., Lubec, G., Pollak, D.D., 2013. Flotillin-1 is an evolutionary-conserved memory-related protein up-regulated in implicit and explicit learning paradigms. Ann. Med. 45 (4), 301–307.

Moret, C., Grimaldi, B., Massot, O., Fillion, G., 2003. The role and therapeutic potential of 5-HT-moduline in psychiatry. Semin Clin Neuropsychiatr. 8 (2), 137–146.

Morris, R.G., 2013. NMDA receptors and memory encoding. Neuropharmacology. 74, 32–40.

Myhrer, T., 2003. Neurotransmitter systems involved in learning and memory in the rat: a meta-analysis based on studies of four behavioral tasks. Brain Res. Rev. 41, 268–287.

Meneses, A., Perez-Garcia, G., 2007. 5-HT$_{1A}$ receptors and memory. Neurosci. Biobehav. Rev. 31 (5), 705–727.

Na, C.H., Jones, D.R., Yang, Y., Wang, X., Xu, Y., Peng, J., 2012. Synaptic protein ubiquitination in rat brain revealed by antibody-based ubiquitome analysis. J. Proteome Res. 11 (9), 4722–4732.

Niciu, M.J., Kelmendic, B., Sanacora, G., 2012. Overview of glutamatergic neurotransmission in the nervous system. Pharmacol. Biochem. Behav. 100, 656–664.

Nonkes, L.J., Maes, J.H., Homberg, J.R., 2013. Improved cognitive flexibility in serotonin transporter knockout rats is unchanged following chronic cocaine self-administration. Addict. Biol. 18 (3), 434–440.

Nordquist, N., Oreland, L., 2010. Serotonin, genetic variability, behaviour, and psychiatric disorders—a review. Ups. J. Med. Sci. 115 (1), 2–10.

Noristani, H.N., Verkhratsky, A., Rodríguez, J.J., 2012. High tryptophan diet reduces CA1 intraneuronal β-amyloid in the triple transgenic mouse model of Alzheimer's disease. Aging Cell. 11 (5), 810–822.

Ögren, S.O., Eriksson, T.M., Elvander-Tottie, E., D'Addario, C., Ekström, J.C., Svenningsson, P., et al., 2008. The role of 5-HT$_{1A}$ receptors in learning and memory. Behav. Brain Res. 195, 54–77.

Olsen, R.W., Sieghart, W., 2008. International union of pharmacology. LXX. Subtypes of γ-aminobutyric acid receptors: classification on the basis of subunit composition, pharmacology, and function. Update Pharmacol. Rev. 60 (3), 243–260.

Oscos, A., Martinez Jr, J.L., McGaugh, J.L., 1988. Effects of post-training D-amphetamine on acquisition of an appetitive autoshaped lever press response in rats. Psychopharmacology (Berl.). 95 (1), 132–134.

Paille, V., Fino, E., Du, K., Morera-Herreras, T., Perez, S., Kotaleski, J.H., et al., 2013. GABAergic circuits control spike-timing-dependent plasticity. J. Neurosci. 33 (22), 9353–9363.

Packard, M.G., Knowlton, B.J., 2002. Learning and memory functions of the Basal Ganglia. Annu. Rev. Neurosci. 25, 563–593.

Pattij, T., 2002. "5-HT$_{1A}$ receptor knockout mice and anxiety: Behavioral and physiological studies." Ph.D. thesis, Universiteit Utrecht, The Netherlands.

Peele, D.B., Vincent, A., 1989. Strategies for assessing learning and memory, 1978–1987: a comparison of behavioral toxicology, psychopharmacology, and neurobiology. Neurosci. Biobehav. Rev. 13 (4), 317–322.

Peixoto, L., Abel, T., 2013. The role of histone acetylation in memory formation and cognitive impairments. Neuropsychopharmacology 38 (1), 62–76.

Pennanen, L., van der Hart, M., Yu, L., Tecott, L.H., 2013. Impact of serotonin (5-HT)$_{2C}$ receptors on executive control processes. Neuropsychopharmacology 38 (6), 957–967.

Pérez-García, G., Meneses, A., 2008a. Memory formation, amnesia, improved memory and reversed amnesia: 5-HT role. Behav. Brain Res. 195, 17−29.

Pérez-García, G., Meneses, A., 2008b. *Ex-vivo* study of 5-HT$_{1A}$ and 5-HT$_7$ receptor agonists and antagonists on cAMP accumulation during memory formation and amnesia. Behav. Brain Res. 195, 139−146.

Pérez-García, G., González-Espinosa, C., Meneses, A., 2006. An mRNA expression analysis of stimulation and blockade of 5-HT$_7$ receptors during memory consolidation. Behav. Brain Res. 169 (2006), 83−92.

Perez-Garcia, G., Meneses, A., 2009. Memory time-course: mRNA 5-HT$_{1A}$ and 5-HT$_7$ receptors. Behav. Brain Res. 202, 102−113.

Prado-Alcalá, R.A., Ruiloba, M.I., Rubio, L., Solana-Figueroa, R., Medina, C., Salado-Castillo, R., et al., 2003. Regional infusions of serotonin into the striatum and memory consolidation. Synapse. 47 (3), 169−175.

Puig, M.V., Gulledge, A.T., 2011. Serotonin and prefrontal cortex function: neurons, networks, and circuits. Mol. Neurobiol. 44 (3), 449−464.

Radley, J.J., Farb, C.R., He, Y., Janssen, W.G., Rodrigues, S.M., Johnson, L.R., et al., 2007. Distribution of NMDA and AMPA receptor subunits at thalamo-amygdaloid dendritic spines. Brain Res. 1134, 87−94.

Rahn, E.J., Guzman-Karlsson, M.C., Sweatt, J., 2013. Cellular, molecular, and epigenetic mechanisms in non-associative conditioning: implications for pain and memory. Neurobiol. Learn. Mem. 105, 133−150.

Rajasethupathy, P., Antonov, I., Sheridan, R., Frey, S., Sander, C., Tuschl, T., et al., 2012. A role for neuronal piRNAs in the epigenetic control of memory-related synaptic plasticity. Cell. 149 (3), 693−707.

Ramírez, M.J., 2013. 5-HT$_6$ Receptors and Alzheimer's disease. Alzheimer's Res. Ther. 5, 15−23.

Raymond, J.R., Mukhin, Y.V., Gelasco, A., Turner, J., Collinsworth, G., Gettys, T.W., et al., 2001. Multiplicity of mechanisms of serotonin receptor signal transduction. Pharmacol. Ther. 92 (2−3), 179−212.

Raymond, J.R., Turner, J.H., Gelasco, A.K., Ayiku, H.B., Coaxum, S.D., Arthur, J.M., et al., 2006. In: Roth, B.L. (Ed.), The Serotonin Receptors. Human Press, Totowa, NJ, pp. 143−206.

Reid, M., Carlyle, I., Caulfield, W.L., Clarkson, T.R., Cusick, F., Epemolu, O., et al., 2010. The discovery and SAR of indoline-3-carboxamides—a new series of 5-HT$_6$ antagonists. Bioorg. Med. Chem. Lett. 20 (12), 3713−3716.

Reis, H.J., Guatimosim, C., Paquet, M., Santos, M., Ribeiro, F.M., Kummer, A., et al., 2009. Neuro-transmitters in the central nervous system and their implication in learning and memory processes. Curr. Med. Chem. 16 (7), 796−840.

Roberts, A.J., Hedlund, P.B., 2012. The 5-HT$_7$ receptor in learning and memory. Hippocampus 22 (4), 762−771.

Roberts, P.D., Spiros, A., Geerts, H., 2012. Simulations of symptomatic treatments for Alzheimer's disease: computational analysis of pathology and mechanisms of drug action. Alzheimers Res. Ther. 4 (6), 50 [Epub ahead of print] PMID: 23181523.

Rodríguez, J.J., Noristani, H.N., Verkhratsky, A., 2012. The serotonergic system in ageing and Alzheimer's disease. Prog. Neurobiol. 99, 15−41.

Romero, G., Sánchez, E., Pujol, M., Pérez, P., Codony, X., Holenz, J., et al., 2006. Efficacy of selective 5-HT$_6$ receptor ligands determined by monitoring 5-HT$_6$ receptor-mediated cAMP signaling pathways. Br. J. Pharmacol. 148, 1133−1143.

Roopra, A., Dingledine, R., Hsieh, J., 2012. Epigenetics and epilepsy. Epilepsia. 53 (Suppl. 9), 2−10.

Roozendaal, B., McGaugh, J.L., 2011. Memory modulation. Behav. Neurosci. 125 (6), 797−824.

Rossé, G., Schaffhauser, H., 2010. 5-HT$_6$ Receptor antagonists as potential therapeutics for cognitive impairment. Curr. Top. Med. Chem. 10, 207−221.

Roth, B.L., Hanizavareh, S.M., Blum, A.E., 2004. Serotonin receptors represent highly favorable molecular targets for cognitive enhancement in schizophrenia and other disorders. Psychopharmacology (Berl.). 174, 17−24.

Ruiz N.V., Oranias G.O., 2010. Patents. In: Borsini, F. (Ed.), Int. Rev. Neurobiol. 5-HT6 Receptors, Part I, Elsevier, Academic Press, Oxford, pp. 36−66.

Ruotsalainen, S., Miettinen, R., MacDonald, E., Riekkinen, M., Sirviö, J., 1998. The role of the dorsal raphe-serotonergic system and cholinergic receptors in the modulation of working memory. Neurosci. Biobehav. Rev. 22 (1), 21−31.

Sarter, M., Hagan, J., Dudchenko, P., 1992. Behavioral screening for cognition enhancers: from indiscriminate to valid testing: part I. Psychopharmacology (Berl.). 107 (2−3), 144−159.

Sawyer, J., Eaves, E.L., Heyser, C.J., Maswood, S., 2012. Tropisetron, a 5-HT$_3$ receptor antagonist, enhances object exploration in intact female rats. Behav. Pharmacol. 23 (8), 806−809.

Schechter, L.E., Smith, D.L., Rosenzweig-Lipson, S., Sukoff, S.J., Dawson, L.A., Marquis, K., et al., 2005. Lecozotan (SRA-333): A selective serotonin1A receptor antagonist that enhances the stimulated release of glutamate and acetylcholine in the hippocampus and possesses cognitive-enhancing properties. J. Exp. Pharmacol. Ther. 314, 1274−1289.

Schliebs, R., Arendt, T., 2011. The cholinergic system in aging and neuronal degeneration. Behav. Brain Res. 221 (2), 555−563.

Schreiber, R., Vivian, J., Hedley, L., Szczepanski, K., Secchi, R.L., Zuzow, M, et al., 2007. Effects of the novel 5-HT$_6$ receptor antagonist RO4368554 in rat models for cognition and sensorimotor gating. Eur. Neuropsychopharmacol. 17 (4), 277−288.

Skelton, M.R., Williams, M.T., Vorhees, C.V., 2008. Developmental effects of 3,4-methylenedioxymethamphetamine: a review. Behav. Pharmacol. 19 (2), 91−111.

Sossin, W.S., 2008. Defining memories by their distinct molecular traces. Trends Neurosci. 31, 170−175.

Spiegel, D.R., Alexander, G., 2011. A case of nonfluent aphasia treated successfully with speech therapy and adjunctive mixed amphetamine salts. J. Neuropsychiatry Clin. Neurosci. 23 (1), E24.

Squire, L.R., Davis, H.P., 1981. The pharmacology of memory: a neurobiological perspective. Annu. Rev. Pharmacol. Toxicol. 21, 323−356.

Squire, L.R., Zola, S.M., 1996. Structure and function of declarative and nondeclarative memory systems. Proc. Natl. Acad. Sci. USA. 93 (24), 13515−13522.

Stahlman, W.D., Young, M.E., Blaisdell, A.P., 2010. Response variability in pigeons in a Pavlovian task. Learn. Behav. 38 (2), 111−118.

Steckler, T., Sahgal, A., 1995. The role of serotonergic−cholinergic interactions in the mediation of cognitive behaviour. Behav. Brain Res. 67, 165−199.

Stern, S.A., Alberini, C.M., 2013. Mechanisms of memory enhancement. Wiley. Interdiscip Rev. Syst. Biol. Med. 5 (1), 37−53.

Sweatt, J.D., 2009. Experience-dependent epigenetic modifications in the central nervous system. Biol. Psychiatry 65, 191−197.

Takahashi, H., Yamada, M., Suhara, T., 2012. Functional significance of central D1 receptors in cognition: beyond working memory. J. Cereb. Blood Flow Metab. 32 (7), 1248−1258.

Talpos, J.C., Fletcher, A.C., Circelli, C., Tricklebank, M.D., Dix, S.L., 2012. The pharmacological sensitivity of a touchscreen-based visual discrimination task in the rat using simple and perceptually challenging stimuli. Psychopharmacology (Berl.). 221 (3), 437−449.

Telese, F., Gamliel, A., Skowronska-Krawczyk, D., Garcia-Bassets, I., Rosenfeld, M.G., 2013. "Seq-ing" insights into the epigenetics of neuronal gene regulation. Neuron 77 (4), 606−623.

Tellez, R., Rocha, L., Castillo, C., Meneses, A., 2010. Autoradiographic study of serotonin transporter during memory formation. Behav. Brain Res. 212 (1), 12−26.

Tellez, R., Gómez-Víquez, L., Meneses, A., 2012a. GABA glutamate, dopamine and serotonin transporters expression on memory formation and amnesia. Neurobiol. Learn. Mem. 97 (2), 189−201.

Tellez, R., Gómez-Viquez, L., Liy-Salmeron, G., Meneses, A., 2012b. GABA glutamate, dopamine and serotonin transporters expression on forgetting. Neurobiol. Learn. Mem. 98 (1), 66−77.

Terry Jr., A.V., Buccafusco, J.J., Wilson, C., 2008. Cognitive dysfunction in neuropsychiatric disorders: selected serotonin receptor subtypes as therapeutic targets. Behav. Brain Res. 195, 30−38.

Terry Jr., A.V., Callahan, P.M., Hall, B., Webster, S.J., 2011. Alzheimer's disease and age-related memory decline (preclinical). Pharmacol. Biochem. Behav. 99 (2), 190−210.

Thellier, M., Lüttge, U., 2013. Plant memory: a tentative model. Plant Biol. (Stuttg.). 15 (1), 1−12.

Thompson, A.J., 2013. Recent developments in 5-HT$_3$ receptor pharmacology. Pharmacol. Sci. 34 (2), 100−109.

Timotijević, I., Stanković, Ž., Todorović, M., Marković, S.Z., Kastratović, D.A., 2012. Serotonergic organization of the central nervous system. Psychiatr. Danub. 201 (Suppl. 3), S326−S330.

Tohyama, M., Takatsuji, K., Kantha, S.S. (Eds.), 1998. Atlas of Neuroactive Substances and Their Receptors in the Rat. Oxford University Press, Oxford.

Tomie, A., Lincks, M., Nadarajah, S.D., Pohorecky, L.A., Yu, L., 2012. Pairings of lever and food induce Pavlovian conditioned approach of sign-tracking and goal-tracking in C57BL/6 mice. Behav. Brain Res. 226 (2), 571−578.

Tremblay, M.A., Acker, C.M., Davies, P., 2010. Tau phosphorylated at tyrosine 394 is found in Alzheimer's disease tangles and can be a product of the Abl-related kinase, Arg. J. Alzheimers Dis. 19, 721−733.

Tsuruoka, N., Beppu, Y., Koda, H., Doe, N., Watanabe, H., Abe, K., 2012. A DKP cyclo (L-Phe-L-Phe) found in chicken essence is a dual inhibitor of the serotonin transporter and acetylcholinesterase. PLoS One. 7 (11), e50824, <10.1371/journal.pone.0050824> Epub 2012 Nov 28.

Turner, J.N., Coaxum, S.D., Gelasco, A.K., Garnovskaya, M.N., Raymond, J.R., 2007. Calmodulin is a 5-HT receptor-interacting and regulatory protein. In: Chattopadhyay, A. (Ed.), Serotonin Receptors in Neurobiology. CRC Press, Boca Raton, FL, Chapter 4 Available from: <http://www.ncbi.nlm.nih.gov/books/NBK5202/>.

Upton, N., Chuang, T.T., Hunter, A.J., Virley, D.J., 2008. 5-HT$_6$ Receptor antagonists as novel cognitive enhancing agents for Alzheimer's disease. Neurotherapeutics 5, 458−469.

van Praag, H.M., 2004. The cognitive paradox in posttraumatic stress disorder: a hypothesis. Prog. Neuropsychopharmacol. Biol. Psychiatry 28 (6), 923−935.

Vanover, K.E., Barrett, J.E., 1998. An automated learning and memory model in mice: pharmacological and behavioral evaluation of an autoshaped response. Behav. Pharmacol. 9 (3), 273−283.

Vermetten, E., Lanius, R.A., 2012. Biological and clinical framework for posttraumatic stress disorder, Handbook of Clinical Neurology, vol. 106. pp. 291−342.

Vianna, M.R., Izquierdo, L.A., Barros, D.M., Walz, R., Medina, J.H., Izquierdo, I., 2000. Short- and long-term memory: differential involvement of neurotransmitter systems and signal transduction cascades. An. Acad. Bras. Cienc. 72 (3), 353−364.

Vitalis, T., Ansorge, M.S., Dayer, A.G., 2013. Serotonin homeostasis and serotonin receptors as actors of cortical construction: special attention to the 5-HT$_{3A}$ and 5-HT$_6$ receptor subtypes. Front. Cell Neurosci. 7, 93.

Volk, B., Nagy, B.J., Vas, S., Kostyalik, D., Simig, G., Bagdy, G., 2010. Medicinal chemistry of 5-HT$_{5A}$ receptor ligands: a receptor subtype with unique therapeutical potential. Curr. Top. Med. Chem. 10 (5), 554−578.

Wallace, T.L., Ballard, T.M., Pouzet, B., Riedel, W.J., Wettstein, J.G., 2011. Drug targets for cognitive enhancement in neuropsychiatric disorders. Pharmacol. Biochem. Behav. 99 (2), 130−145.

Ward, B.O., Wilkinson, L.S., Robbins, T.W., Everitt, B.J., 1999. Forebrain serotonin depletion facilitates the acquisition and performance of a conditional visual discrimination task in rats. Behav. Brain Res. 100 (1−2), 51−65.

Wardle, M.C., Yang, A., de Wit, H., 2012. Effect of D-amphetamine on post-error slowing in healthy volunteers. Psychopharmacology (Berl.). 220 (1), 109−115.

Weiner, M.W., Veitch, D.P., Aisen, P.S., Beckett, L.A., Cairns, N.J., Green, R.C., et al., 2012. The Alzheimer's disease neuroimaging initiative: a review of papers published since its inception. Alzheimers Dement. 9 (5), 111−194

Williams, G.V., Rao, S.G., Goldman-Rakic, P.S., 2002. The physiological role of 5-HT$_{2A}$ receptors in working memory. J. Neurosci. 22 (7), 2843−2854.

Wilson, C., Terry, A.V., 2009. Enhancing cognition in neurological disorders: potential usefulness of 5-HT$_6$ antagonists. Drugs Future 34, 969−975.

Woods, S., Clarke, N.N., Layfield, R., Fone, K.C., 2012. 5-HT$_6$ receptor agonists and antagonists enhance learning and memory in a conditioned emotion response paradigm by modulation of cholinergic and glutamatergic mechanisms. Br. J. Pharmacol. 167 (2), 436−449.

Xu, Y., Yan, J., Zhou, P., Li, J., Gao, H., Xia, Y., et al., 2012. Neurotransmitter receptors and cognitive dysfunction in Alzheimer's disease and Parkinson's disease. Prog. Neurobiol. 97 (1), 1−13.

Yamada, K., Nabeshima, T., 2003. Brain-derived neurotrophic factor/TrkB signaling in memory processes. J. Pharmacol. Sci. 91 (4), 267−270.

Youn, J., Misane, I., Eriksson, T.M., Millan, M.J., Ogren, S.O., Verhage, M., et al., 2009. Bidirectional modulation of classical fear conditioning in mice by 5-HT$_{1A}$ receptor ligands with contrasting intrinsic activities. Neuropharmacology 57 (5−6), 567−576.

Yu, H.Y., 2012. The prescription drug abuse epidemic. Clin. Lab. Med. 32 (3), 361−377.

Yun, H.M., Rhim, H., 2011a. 5-HT$_6$ Receptor ligands, EMD386088 and SB258585, differentially regulate 5-HT$_6$ receptor-independent events. Toxicol. In Vitro. 25 (8), 2035–2040.

Yun, H.M., Rhim, H., 2011b. The serotonin-6 receptor as a novel therapeutic target. Exp. Neurobiol. 20 (4), 159–168.

Yun, H.M., Kim, S., Kim, H.J., Kostenis, E., Kim, J.I., Seong, J.Y., et al., 2007. The novel cellular mechanism of human 5-HT$_6$ receptor through an interaction with Fyn. J. Biol. Chem. 282, 5496–5505.

Yun, H.M., Baik, J.H., Kang, I., Jin, C., Rhim, H., 2010. Physical interaction of Jab1 with human serotonin 6G-protein-coupled receptor and their possible roles in cell survival. J. Biol. Chem. 285, 10016–10029.

Zhang, G., Asgeirsdóttir, H.N., Cohen, S.J., Munchow, A.H., Barrera, M.P., Stackman, R. W., 2013. Current neuroimaging techniques in Alzheimer's disease and applications in animal models. Stimulation of serotonin 2A receptors facilitates consolidation and extinction of fear memory in C57BL/6J mice. Neuropharmacology 64, 403–413.

Zhang, L., Chang, R.C., Chu, L.W., Mak, H.K., 2012. Current neuroimaging techniques in Alzheimer's disease and applications in animal models. Am. J. Nucl. Med. Mol. Imaging 2 (3), 386–404.

Zhou, W., Chen, L., Paul, J., Yang, S., Li, F., Sampson, K., et al., 2012. The effects of glycogen synthase kinase-3beta in serotonin neurons. PLoS One 7 (8), e43262.

Zola-Morgan, S., Squire, L.R., 1993. Neuroanatomy of memory. Ann. Rev. Neurosci. 168, 547–563.

3 The Role of GABA in Memory Processes

Antonella Gasbarri and Assunta Pompili

Department of Applied Clinical and Biotechnologic Sciences, University of L'Aquila, Italy

Introduction

The γ-aminobutyric acid (GABA) is an amino acid transmitter synthesized by decarboxylation of glutamate by the enzyme glutamic acid decarboxylase (GAD) (Rowley et al., 2012), which exists in two isoforms, GAD65 and GAD67, having different molecular weights (65 and 67 KDa), catalytic and kinetic properties, and subcellular localization (Walls et al., 2010, 2011).

GABA had been long known to exist in plants and bacteria, where it serves a metabolic role in the Krebs cycle. In the mammalian central nervous system (CNS), there was an extraordinary amount of GABA—1 mg/g—and GABA was virtually undetectable in other tissues (Jorgensen, 2005; for historical perspectives, see Roberts (2000) and Florey (1991)). However, GABA was not accepted as neurotransmitter until the 1960s after extensive physiological experimentation.

GABA represents one of the few amino acids existing in high concentrations in the CNS, which plays a major neurotransmitter role. In particular, GABA is the predominant inhibitory neurotransmitter in the brain (Hassel and Dingledine, 2006; Olsen and Betz, 2006; Schousboe and Waagepetersen, 2008) which acts through two different receptors: $GABA_A$ and $GABA_B$. (Sigel and Steinmann, 2012) The $GABA_A$ receptors represent the major inhibitory neurotransmitter receptors in mammalian brain. $GABA_B$ are G-protein coupled receptors (GPCRs) and differ strongly in structure, function, and sequence from $GABA_A$ receptors. $GABA_C$ receptors are now generally assumed to be one of the many isoforms of $GABA_A$ receptors, and the International Union of Basic and Clinical Pharmacology discourages further use of this term (Olsen and Sieghart, 2008).

$GABA_A$ receptors are the principal sites of fast synaptic inhibition and are involved in the regulation of vigilance, anxiety, muscle tension, epileptogenic activity, and memory functions (Korpi et al., 2002; Rudolph and Möhler, 2004).

$GABA_A$ receptors are responsive to a wide variety of drugs, e.g., BZs, which are often used for their sedative/hypnotic and anxiolytic effects (Rudolph et al., 2001; Sieghart and Sperk, 2002).

Identification of Neural Markers Accompanying Memory. DOI: http://dx.doi.org/10.1016/B978-0-12-408139-0.00003-1

GABA$_B$ receptors modulate behavior and cerebral reward processes (Vlachou and Markou, 2010). GABA$_B$ agonists and positive modulators have been found to inhibit the reinforcing effects of addictive drugs, such as cocaine, amphetamine, opiates, ethanol, and nicotine. The converging evidence of the effects of GABA$_B$ agonists and positive modulators on the reinforcing properties of drugs of abuse is based on the behavioral studies utilizing a variety of procedures with relevance to reward processes and drug abuse liability (such as intracranial self-stimulation, intravenous self-administration under both fixed- and progressive-ratio schedules of reinforcement, and conditioned place preference). In animal models, GABA$_B$ agonists and positive modulators block the reinforcing effects of drugs of abuse. However, GABA$_B$ receptor agonists also cause side effects. When utilized as medications for drug addiction, GABA$_B$ receptor modulators have potential advantages. They have a better side effect profile compared to GABA$_B$ agonists because, in the absence of GABA, they are devoid of intrinsic agonistic activity. They only exert their modulatory actions in concert with endogenous GABAergic activity (Vlachou and Markou, 2010). Therefore, GABA$_B$ positive modulators are promising therapeutics for the treatment of various aspects of dependence (e.g., initiation, maintenance, and relapse) on various drugs of abuse, such as cocaine, heroin, alcohol, and nicotine.

GABA Receptors

GABA exerts its inhibitory role through two types of receptors, ionotropic and metabotropic (Goudet et al., 2009; Hassel and Dingledine, 2006; Olsen and Betz, 2006; Savić et al., 2005c; Schousboe and Waagepetersen, 2008). While ionotropic receptors are ligand gated ion channels involved in fast synaptic transmission, metabotropic receptors belong to the superfamily of GPCRs and are responsible for the neuromodulatory effect of GABA. Ionotropic receptors include GABA$_A$, which are more abundant than GABA$_B$ within the brain (Pirker et al., 2000); metabotropic receptors, widely distributed throughout the brain, include GABA$_B$; GABA$_C$ receptors are now assumed to be one of the several isoforms of GABA$_A$ receptors.

GABA$_A$ receptors, which share their structural characteristics with the entire superfamily of Cys loop-type ligand-gated ion channels, are heteropentameric (generally pentameric) chloride-selective ligand-gated ion channels (Jurd and Moss, 2010; Sigel and Steinmann, 2012).

Binding of GABA to this receptor triggers opening of the channel, causing influx of negatively charged chloride ions into the neuron, causing reduced excitatory neurotransmission. The mammalian GABA$_A$ receptor is composed of seven classes of subunits, each having multiple variants ($\alpha1-\alpha6$, $\beta1-\beta3$, $\gamma1-\gamma3$, $\rho1-\rho3$, δ, ϵ, θ) (Pirker et al., 2000; Rudolph et al., 2001). Most functional GABA$_A$ receptors are made up of two α-subunits, two β-subunits, and one γ-subunit or alternatively two α-subunits, one β-subunit, and two γ-subunits, which together comprise the central ion channel (Esmaeili et al., 2009; Haefely, 1989; Mehta and

Ticku, 1999; Pirker et al., 2000; Savić et al., 2005c). In the mammalian CNS, the most predominant GABA$_A$ receptors have an α1 β2 γ2 combination (Fritschy et al., 1992).

There are several different ligands binding to GABA$_A$ receptors, many of which have distinct binding sites. GABA$_A$ receptor agonists include full agonists such as GABA and muscimol, which bind to and activate the GABA$_A$ receptor complex at the GABA-binding site, at the interface of α- and β-subunits (Mehta and Ticku, 1999). As a consequence, chloride channels open, leading to an influx of chloride ions and increase of neuronal inhibition (Johnston, 1996). Other GABA$_A$ receptor agonists include benzodiazepines (BZs) (e.g., midazolam) binding to a separate site at the interface of the α - and γ-subunits (Mehta and Ticku, 1999). Barbiturates (e.g., pentobarbital) and neurosteroids (e.g., allopregnanolone) represent other types of agonists binding to the GABA$_A$ complex at a distinct site from both GABA and BZs (Amin and Weiss, 1993; Mehta and Ticku, 1999). They both potentiate GABAergic responses at small doses but, at higher concentrations, may activate the receptor directly (Mehta and Ticku, 1999). GABA$_A$ receptor partial agonists not only show similar effects to the full agonists, but also show a decreased efficacy of binding to and activating the GABA$_A$ receptor complex (Johnston, 1996).

The GABA$_A$ receptor antagonists represent an other family of compounds associated with GABA$_A$ receptors. Competitive antagonists (such as bicuculline) occupy the GABA-binding site, then preventing GABA from binding to and activating the receptor. It is important to note that these antagonists may affect behavior in the presence of a tonic GABAergic inhibition. Noncompetitive antagonists (such as picrotoxin) antagonize the inhibitory effects of GABA by binding to picrotoxin-binding sites localized at the chloride ion channel of GABA$_A$ receptors that could induce the closure of the chloride ion channel. This action, blocking the movement of chloride ions into the channel, prevents hyperpolarization and, as a consequence, reduces inhibitory transmission (Johnston, 1996). Antagonists such as flumazenil bind to the BZ site, blocking the access of agonists, and inverse agonists to this binding site. However, these compounds do not prevent GABA (and other direct agonists or antagonists) to bind its binding site. The last family of GABA$_A$ receptor ligands are the inverse agonists. Full inverse agonists bind to the BZs site but decrease inhibitory GABA transmission by decreasing both the chloride channel opening and the affinity for GABA to bind to and activate GABA$_A$ receptors (Johnston, 1996). Partial inverse agonists are similar to the inverse agonists but show a decreased efficacy of binding to and inducing a functional change in the receptor (Harris and Westbrook, 1998).

They are composed of different subunits that may be well characterized concerning sequence, expression level, and localization in a neuron. In the CNS, there are at least 19 subunits. Nevertheless, the vast majority of receptors appear to be an association of two α-subunits, two β-subunits, and a single γ-subunit, which make up a central ion channel. Most of them contain a BZ binding site located at the interface of the γ-subunit and the respective α-subunit (α1, α2, α3, or α5) (Korpi et al., 2002). Research using genetically modified mice has pointed to the specific contribution of specific receptor subtypes to the pharmacologic spectrum of BZs.

In particular, the sedative and anterograde amnesic effects of BZs were mainly attributed to $\alpha 1$-containing $GABA_A$ receptor subtypes (Savić et al., 2005c). $GABA_A$ mediate fast synaptic inhibition in the adult CNS and represent the major inhibitory neurotransmitter receptors in mammalian brain.

$GABA_A$ receptors, located in the postsynaptic membrane, mediate neuronal inhibition occurring in the millisecond time range, while those located in the extrasynaptic membrane respond to ambient GABA and confer long-term inhibition.

The agonist of $GABA_A$ receptors is generally being called after GABA, in spite of the fact that, under physiological conditions, the acid form of the neurotransmitter hardly exists, and a more appropriate name for this neurotransmitter would be γ-aminobutyrate.

Regulation of the number of $GABA_A$ receptors at synapses is critical to maintain the correct level of synaptic inhibitory transmission and physiological function. Phosphorylation of $GABA_A$ receptor subunits represents one such mechanism leading to the dynamic modulation of $GABA_A$ receptor function.

$GABA_A$ receptors are involved in the regulation of vigilance, anxiety, muscle tension, epileptogenic activity, and memory functions (Korpi et al., 2002; Rudolph and Möhler, 2004). They are responsive to a wide variety of drugs, e.g., BZs, which are often used for their sedative/hypnotic and anxiolytic effects.

Metabotropic GABA receptors, widely distributed throughout the brain, include $GABA_B$, which are GPCRs differing strongly in structure, function, and sequence from $GABA_A$ receptors. They belong to the superfamily of GPCRs, constituting a large family of membrane proteins responsible for transduction of various external signals into intracellular responses through heterotrimeric G-proteins. As a result, GPCRs are involved in the regulation of several physiological and pathological processes and are the target of about a quarter of drugs available on the market (Goudet et al., 2009; Overington et al., 2006). Based on sequence comparison, five main classes of GPCRs, sharing no sequence similarity, have been identified. However, all these receptors have in common a central core domain composed of seven transmembrane helices which is responsible for G-protein coupling (Bockaert and Pin, 1999). $GABA_B$ receptors belong to class C GPCRs (formerly known as family 3 GPCRs), which also includes the calcium-sensing receptor, the receptors for sweet and umami taste and different pheromone and orphan receptors (Brauner-Osborne et al., 2007). Class C GPCRs has two important particularities that are relevant for their regulations and function. The first structural characteristic of most class C receptors (except pheromone receptors) is the presence of a large bilobate extracellular domain where natural ligands bind. This domain is juxtaposed to the core transmembrane domain common to all GPCRs and responsible of G-protein coupling. The second specific feature of class C GPCRs is represented by their dimeric nature (Pin et al., 2004b). The transmembrane domain is important for the pharmacology of class C GPCRs (Pin et al., 2004b). In the late 1990s, a new class C GPCRs ligands was identified, the allosteric $GABA_B$ is a heterodimeric receptor, composed of two subunits, $GABA_{B1}$ and $GABA_{B2}$. $GABA_{B1}$ is responsible for ligand recognition but is unable to reach the cell surface by itself or to activate G-proteins (Filippov et al., 2000; Galvez et al., 2001; Margeta-Mitrovic et al., 2001a,b). On the contrary, $GABA_{B2}$ is unable to bind GABA but is responsible for G-protein coupling (Duthey et al., 2002; Havlickova et al., 2002; Kniazeff et al., 2002). In addition, $GABA_{B2}$ masks the retention signal located on the C terminus of $GABA_{B1}$ and allows the expression of the heterodimers at the cell surface (Brock et al., 2005). Therefore, to get a functional receptor, $GABA_{B1}$ and $GABA_{B2}$ need to be associated (Pin et al., 2004a).

$GABA_B$ receptors modulation plays a central role in the capability of neurons to function in circuits (Chalifoux and Carter, 2011).

This is highlighted by the consequences of disrupted modulation in the prefrontal cortex in neuropsychiatric diseases (Bowery, 2006). Recent studies have revealed new ways in which $GABA_B$ receptors can control synaptic responses. Thus, they can suppress multivesicular release, in response to a single action potential, to decrease the synaptic glutamate concentration. Unexpectedly, $GABA_B$ receptors can also act via the protein kinase A (PKA) pathway to decrease postsynaptic N-methyl-D-aspartate (NMDA) receptors Ca^{2+} signals. Therefore, by also inhibiting voltage-sensitive Ca^{2+} channels in spines and dendrites, $GABA_B$ receptors are poised to strongly regulate Ca^{2+}-mediated plasticity. In addition to $GABA_B$ receptors, these effects are also found with other neurotransmitters like acetylcholine and dopamine, suggesting that these processes are occurring at different synapses throughout the brain. However, many questions remain open concerning the spatial, temporal, and cell-type specific effects of neuromodulators. A variety of new technologies will allow to clarify the characteristics of synaptic modulation in normal physiology and disease states.

$GABA_C$ receptors are now assumed to be one of the several isoforms of $GABA_A$ receptors, and the International Union of Basic and Clinical pharmacology discourages further use of this term (Olsen and Sieghart, 2008; Sigel and Steinmann, 2012).

GABA and Its Correlation with Memory

The involvement of GABA in the regulation of vigilance, anxiety, and memory processes is well established (Korpi et al., 2002; Rudolph and Möhler, 2004).

Pharmacological studies have demonstrated that posttraining injections of GABAergic compounds modulate memory storage (Hatfield et al., 1999). These findings strongly support the view that $GABA_A$ receptors modulate posttraining processes underlying memory consolidation (Brioni and McGaugh, 1988; Castellano and McGaugh, 1990). It was also reported that bicuculline causes memory facilitation when infused into CA1 either immediately after training or 1.5 h posttraining (Luft et al., 2004). Similarly, bicuculline improves memory consolidation in an invertebrate model using the crab *Chasmagnathus* (Carbo Tano et al., 2009). Consistent with these previous studies, Kim et al. (2012) reported that bicuculline methiodide, when systemically administered immediately or 1 h after the acquisition trial (but not 3 h later), enhances memory consolidation in the one-trial passive avoidance task, suggesting that $GABA_A$ receptor blockade enhances memory consolidation.

Besides the site of action of GABA itself, several modulatory sites at $GABA_A$ receptors exist, mediating the actions of many drugs, such as BZs (Chebib and Johnston, 2000; Korpi et al., 2002). The characterization of the various pharmacological effects of the BZs (sedative, anxiolytic, muscle relaxant, amnesic, etc.) is considered a major success of behavioral pharmacology (Sanger et al., 2003). Since the introduction of chlordiazepoxide in 1960, BZs have been extensively prescribed to cope with anxiety, insomnia, muscle spasm, and epilepsy and they are still considered one of the best pharmacological treatment for anxiety disorders (Sramek et al., 2002), their side effects, including impairment of mnesic function (Savić et al., 2005c), have to be taken into account.

GABA also play an important role in spatial memory, interacting with NMDA receptors.

GABA, Memory and BZ

It is well known that $GABA_A$ receptor agonists impair memory function and that its antagonists enhance memory consolidation (McGaugh and Roozendaal, 2009). For example, $GABA_A$ receptor antagonists, including bicuculline and flumazenil, enhance performance in memory tasks (Herzog et al., 1996), and its agonists, such as muscimol and diazepam, impair memory formation (Castellano and McGaugh, 1989). In particular, it was demonstrated that bicuculline increases memory consolidation when administered into the hippocampal CA1 region either immediately or 1.5 h after training (Luft et al., 2004). In addition, $GABA_A$ receptor antagonists (including bicuculline) enhance brain-derived neurotrophic factor (BDNF) expression in the hippocampal formation (HF) (Katoh-Semba et al., 2001; Metsis et al., 1993).

If BDNF is generally used to improve memory consolidation, enhanced memory consolidation by $GABA_A$ receptor blockade retrieved 24 h after an acquisition trial might depend on the increased BDNF levels (Bekinschtein et al., 2007). However, it is unclear which elements or signaling pathways are involved in the enhancement of memory consolidation by $GABA_A$ receptors blockade.

Kim et al. (2012) hypothesized that the enhancement of memory consolidation, caused by blockade of $GABA_A$ receptors within a limited time window, depend on the increase of BDNF levels, induced by $GABA_A$ receptor blockade. The results of their study suggest that the enhancement of the level of mBDNF and its function during a restricted time window after training are required for the enhancement of memory consolidation induced by $GABA_A$ receptor blockade.

BZs have been repeatedly associated to the impairment of memory acquisition and anterograde amnesia. Concerning the retrograde memory effects, the results of most animal studies rule out a role of BZs on retrieval in different memory tasks (McNamara and Skelton, 1991; Venault et al., 1986). However, the results of experiments, including tests that contained a significant emotional component, pointed to the inhibitory (Cole and Michaleski, 1984; Cole, 1986) as well as the facilitative (Obradovic et al., 2004; Savić et al., 2003) influences of the BZs on memory retrieval. In fact, according to the classic Yerkes–Dodson hypothesis (1908), one can expect a curvilinear relation between arousal and/or anxiety and performance, such that a moderate level of anxiety can improve cognitive performance, depending on task difficulty (Eysenck, 1985).

Similar to findings on animals, reports of human studies suggest that BZs do not significantly influence memory retrieval (Ghoneim and Mewaldt, 1975; Lister, 1985). However, retrieval impairment in young adult males has been reported (Block and Berchou, 1984), and memory improvement in humans is sometimes evidenced (File et al., 1999; Fillmore et al., 2001; Hinrichs et al., 1984). It has been hypothesized that this phenomenon is not a true facilitation of retrieval processes but could be the consequence of reduced interference from items presented after drug administration as a paradoxical effect of drug-induced anterograde amnesia (Hinrichs et al., 1984). However, facilitating effects on retrieval processes that are more specific have also been proposed (File et al., 1999).

A subtype specificity of $GABA_A$ receptors in the effects of BZs on memory has been evidenced (Savić et al., 2005c). In fact, explicit memory seems to be influenced by the $GABA_A$ receptors containing the $\alpha 1$- and α $\alpha 5$-subunits, whereas the effects on procedural memory can be mainly mediated by the $\alpha 1$-subunit. The involvement of the $\alpha 1$-subunit in memory modulation is not unexpected, considering that it is the major subtype, present in 60% of all $GABA_A$ receptors. On the other hand, the role of $\alpha 5$-subunits, mainly expressed in the HF, in modulating distinct aspects of memory gives promise of selective pharmacological coping with specific memory deficits. The results of experiments using the preferential $\alpha 1$-subunit selective antagonist β-CCt suggest that the inhibitory effects of BZs agonists on the formation of explicit memory, in a passive avoidance test, involve other α-subunits in addition to the $\alpha 1$ subtype (Savić et al., 2005a). Hippocampal $GABA_A$ circuits expressing $\alpha 5$-subunit may be critically involved in the memory for this particular task (Möhler et al., 2004). On the other hand, the $\alpha 1$-subunit is substantially involved in procedural memory processing (Savić et al., 2005b). The important role of the $\alpha 1$-subunit in memory modulation is not unexpected, considering that this subunit represent the major subtype found in 60% of all $GABA_A$ receptors (Möhler et al., 2002). On the other hand, the role of $\alpha 5$-subunits, mainly expressed in the HF, in modulating distinct forms of memory, gives promise of specific pharmacological coping with specific memory deficits through the selective inverse agonism in this receptor subpopulation (Street et al., 2004).

GABA and Spatial Memory

It is well known that hippocampal $GABA_A$ and NMDA receptors play an important role in spatial memory. Intrahippocampal injection of the $GABA_A$ receptor agonist muscimol impaired retrieval in the water maze task in rats (Moser and Moser, 1998). It was also reported that posttraining intra-HF injection of muscimol resulted in short-term memory impairment in object recognition (de Lima et al., 2006). Intrahippocampal injection of (+)MK-801 or 2-amino-5-phosphonopentanoic acid (AP5), an NMDA receptor antagonist, also caused spatial memory deficits in radial arm maze performance in rats (Houston et al., 2007; Houston and Smart, 2006; Huang et al., 2003; 2004; Yoshihara and Ichitani, 2004). In contrast, posttraining intrahippocampal injection of AP5 showed no impairment of the retention of working memory in the delay-interposed radial maze (Yoshihara and Ichitani, 2004) reported that. The discrepancy between the above results might be due to the difference in the measurement method for spatial memory.

On the other hand, *in vitro* studies evidenced that hippocampal $GABA_A$ and NMDA receptors have a close interaction. For example, the activation of NMDA receptor by NMDA or glutamate application caused suppression of the $GABA_A$ receptor response in HF pyramidal neurons and cerebellar granule cells; this effect was mediated by the activation of calcineurin through NMDA receptor (Chen and Wong, 1995; Lu et al., 2000; Robello et al., 1997; Stelzer and Shi, 1994). Therefore, *in vivo* studies evidenced that NMDA receptor may also interact with $GABA_A$ receptor in the retention of working memory.

The effects of drugs on the retention of spatial working memory can be effectively investigated using the delayed spatial win-shift (SWSh) task (Yoshihara and Ichitani, 2004); therefore, using this method, Saito et al. (2010) first examined how long rats can store and utilize information about their own responses to efficiently obtain a reward located in the maze. Next, in order to clarify the interaction between $GABA_A$ and NMDA receptors in the retention of working memory, the effects of muscimol, (+)MK-801 and the combined use of these drugs on a delayed SWSh task in a radial arm maze were studied. A small rise in Ca^{2+} through the NMDA receptor channel induced preferential activation of calcineurin reported (Malenka, 1994). As shown previously, the results of *in vitro* studies showed the participation of calcineurin in the response of $GABA_A$ receptor. Therefore, in order to clarify the involvement of the calcineurin signaling pathway in the interaction between $GABA_A$ and NMDA receptors, Saito et al. (2010) investigated the effects of cyclosporin A, a calcineurin inhibitor, and the combined use of muscimol and cyclosporin A, on the retention of a delayed SWSh task in a radial arm maze. The results of these studies indicated that hippocampal NMDA receptors regulate the effect of spatial working memory induced by muscimol. In addition, the calcineurin signaling pathway may be involved in muscimol-induced impairment of memory retention.

While the pharmacological manipulation of $GABA_A$ receptors function by therapeutic agents, such as BZs can have profound action on neuronal excitation and behavior, the endogenous mechanisms utilized by neurons to regulate the efficacy of synaptic inhibition and their impact on behavior is still poorly understood. To address this issue, Tretter et al. (2009) created a knock-in mouse in which tyrosine phosphorylation of the $GABA_A$ $\gamma 2$ subunit, a posttranslational modification that is crucial for their functional modulation, was ablated. The authors choose to focus their studies on the HF, considering its well-established role in learning and memory. These knock-in mice exhibited enhanced $GABA_A$ receptors accumulation at postsynaptic inhibitory synaptic specializations on pyramidal neurons within the CA3 subdomain of the HF, mainly caused by aberrant trafficking within the endocytic pathway. This increased inhibition correlated with a specific deficit in spatial object recognition, a behavioral paradigm dependent upon CA3. Thus, phospho-dependent modulation of $GABA_A$ receptors function involving just two tyrosine residues in the γ2 subunit provides an input-specific mechanism that not only regulates the efficacy of synaptic inhibition but also has behavioral consequences.

GABA and Fear Memory

A substantial body of research over the past 30 years has been focused on the psychological processes involved in learned fear and identifying their neural mechanisms. Many lines of evidence have led to the view that fear memory formation, reconsolidation, and extinction depend on reduced activation of $GABA_A$ receptors in different cerebral regions. For example, systemic administration of drugs (e.g., BZs) that activate these receptors alleviates symptoms of anxiety in humans, and their intra-amygdala infusion reduces learned fear responses in non-human animals. This suggests that $GABA_A$ receptors, possibly within the amygdala and HF, have an important role in the acquisition and consolidation of fear memories (Makkar et al., 2010). In agreement with these findings, administration of $GABA_A$ receptor agonists immediately after a brief conditioned stimulus re-exposure disrupt, whereas $GABA_A$ receptor antagonists facilitate, subsequent fear responding, indicating that $GABA_A$ receptors are also involved in the reconsolidation of fear memories after retrieval. Finally, increasing GABAergic transmission both before and immediately after extinction training blocks response inhibition, indicating that activation of $GABA_A$ receptors interferes with the acquisition and consolidation of fear memories. Even though contradictory results have emerged, due to variations in the location of drug infusion, dosage of the drug and/or the time point of drug administration, many evidence strongly suggest that the processes mediating memory persistence after initial fear learning, memory reactivation, and extinction training dependent on a common mechanism of reduced GABAergic transmission (Makkar et al., 2010). It was suggested that this downregulation of GABAergic transmission most likely takes place in the amygdala, HF (e.g., if the CS is a context), or the medial prefrontal cortex (during extinction training). The downregulation of GABA modulates memory storage by facilitating the release of norepinephrine which, after binding to β-ARs, initiates an intracellular cascade culminating in the synthesis of new proteins, which are utilized for the synaptic changes required to stabilize the new, reactivated, or inhibitory memory trace (Makkar et al., 2010).

The evidence suggesting that GABAergic transmission is detrimental to the persistence of fear memories may have implications for the treatment of anxiety disorders in humans, particularly those associated with maladaptive and intrusive fear memories such as posttraumatic stress disorder, social phobia, and specific phobia (Day et al., 2004; Durand and Barlow, 2006; Ehlers and Clark, 2000; Hackmann and Holmes, 2004; Hackmann et al., 2000; Rachman, 1991). In particular, in order to block the reconsolidation of fear memories and, then, reduce subsequent anxiety symptoms, GABA agonists (particularly BZs, such as midazolam or diazepam) could be administered immediately after shortly re-exposing patients to fear-related stimuli. Numerous animal studies have shown that the reduced fear responding produced by midazolam does not recover over time, with a shift in the internal state, or following re-exposure to the unconditioned stimulus (Bustos et al., 2006; 2009; Makkar et al., 2010; Zhang and Cranney, 2008). These findings suggest that combining BZs with short cue exposure might be a useful and lasting treatment for anxiety disorders, taking into account that data exist showing that reconsolidation of conditioned fear can be disrupted in humans by the use of the β-adrenoceptor antagonist propranolol (Miller et al., 2004), and midazolam is already being used in clinical settings for its sedative and anxiolytic effects (Pain et al., 2002). However, if re-exposure to conditioned stimulus is too long, $GABA_A$ agonists could disrupt extinction of the fear memory, as demonstrated by Bustos et al. (2009), having as a consequence a persistence of anxiety symptoms and, then, a worsening of the problem. Therefore, if BZs are to be used in conjunction with brief cue exposure, clinicians will need to ensure that cue exposure is brief, and that decrease of anxiety (i.e., within-session extinction) does not occur throughout the exposure session, to avoid that the BZ will disrupt the extinction memory, leading to maintenance of fear and anxiety. Further studies aiming to clarify the effect of administering $GABA_A$ receptor agonists following variations in the duration of CS re-exposure are required to determine the optimal duration. Moreover, data showing that reconsolidation of older fear memories can be disrupted but require longer conditioned stimulus re-exposures and higher drugs dosages (Bustos et al., 2009) add further complications. In particular, it could be necessary that the clinicians vary the length of re-exposure and drug dosage based on the age of the memory or the duration of the disorder. This may imply a process of trial and error, which could be time consuming for the physician and detrimental to the patient, who does not receive the immediate treatment he/she requires.

However, the finding that midazolam can disrupt the reconsolidation of remote fear memories is promising for the use of BZs and brief cue exposure in treating anxiety disorders, because patients suffering of anxiety disorders often wait several years before seeking treatment (Durand and Barlow, 2006; Foa et al., 2000). Nonetheless, it is difficult to assess the age of fear memories in the context of human anxiety disorders. For example, in specific phobias, such as aracnophobia, patients often do not remember a specific incident that elicited their fear, claiming to have always been afraid of spiders (Rachman, 1991).

Then, the effects of BZs in memory reconsolidation in humans and the chemical efficacy of GABA antagonists facilitating long-duration cue exposure (extinction training) require further investigation.

References

Amin, J., Weiss, D.S., 1993. $GABA_A$ receptor needs two homologous domains of the b subunit for activation by GABA but not by pentobarbital. Nature 366, 565−569.
Bekinschtein, P., Cammarota, M., Igaz, L.M., Bevilaqua, L.R., Izquierdo, I., Medina, J.H., 2007. Persistence of long-term memory storage requires a late protein synthesis- and BDNF-dependent phase in the hippocampus. Neuron 53, 261−277.

Block, R.I., Berchou, R., 1984. Alprazolam and lorazepam effects on memory acquisition and retrieval processes. Pharmacol. Biochem. Behav. 20, 233–241.

Bockaert, J., Pin, J.P., 1999. Molecular tinkering of G protein-coupled receptors: an evolutionary success. EMBO J. 18, 1723–1729.

Bowery, N.G., 2006. GABA_B receptor: a site of therapeutic benefit. Curr. Opin. Pharmacol. 6, 37–43.

Brauner-Osborne, H., Wellendorph, P., Jensen, A.A., 2007. Structure, pharmacology and therapeutic prospects of family C G-protein coupled receptors. Curr. Drug Targets 8, 169–184.

Brioni, J.D., McGaugh, J.L., 1988. Post-training administration of GABAergic antagonists enhances retention of aversively motivated tasks. Psychopharmacology (Berl.) 96, 505–510.

Brock, C., Boudier, L., Maurel, D., Blahos, J., Pin, J.P., 2005. Assembly-dependent surface targeting of the heterodimeric GABA_B receptor is controlled by COPI but not 14-3-3. Mol. Biol. Cell 16, 5572–5578.

Bustos, S.G., Maldonado, H., Molina, V.A., 2006. Midazolam disrupts fear memory reconsolidation. Neuroscience 139, 831–842.

Bustos, S.G., Maldonado, H., Molina, V.A., 2009. Disruptive effect of midazolam on fear memory reconsolidation: decisive influence of reactivation time span and memory age. Neuropsychopharmacology 34, 446–457.

Carbo Tano, M., Molina, V.A., Maldonado, H., Pedreira, M.E., 2009. Memory consolidation and reconsolidation in an invertebrate model: the role of the GABAergic system. Neuroscience 158, 387–401.

Castellano, C., McGaugh, J.L., 1989. Retention enhancement with post-training picrotoxin: lack of state dependency. Behav. Neural. Biol. 51, 165–170.

Castellano, C., McGaugh, J.L., 1990. Effects of post-training bicuculline and muscimol on retention: lack of state dependency. Behav. Neural. Biol. 54, 156–164.

Chalifoux, J.R., Carter, A.G., 2011. GABA_B receptor modulation of synaptic function. Curr. Opin. Neurobiol. 21, 339–344.

Chebib, M., Johnston, G.A., 2000. GABA-activated ligand gated ion channels: medicinal chemistry and molecular biology. J. Med. Chem. 43, 1427–1447.

Chen, Q.X., Wong, R.K., 1995. Suppression of GABA_A receptor responses by NMDA application in hippocampal neurones acutely isolated from the adult guinea-pig. J. Physiol. (Lond.). 482, 353–362.

Cole, S.O., 1986. Effects of benzodiazepines on acquisition and performance: a critical assessment. Neurosci. Biobehav. Rev. 10, 265–272.

Cole, S.O., Michaleski, A., 1984. Chlordiazepoxide impairs the performance of a learned discrimination. Behav. Neural. Biol. 41, 223–230.

Day, S., Holmes, E.A., Hackmann, A., 2004. Occurrence of imagery and its link with early memories in agoraphobia. Memory 12, 416–427.

de Lima, M.N., Luft, T., Roesler, R., Schroder, N., 2006. Temporary inactivation reveals an essential role of the dorsal hippocampus in consolidation of object recognition memory. Neurosci. Lett. 405, 142–146.

Durand, V.M., Barlow, D.H., 2006. Essentials of Abnormal Psychology. fourth ed. Thomson-Wadsworth, Belmont.

Duthey, B., Caudron, S., Perroy, J., Bettler, B., Fagni, L., Pin, J.P., et al., 2002. A single subunit (GB2) is required for G-protein activation by the heterodimeric GABA(B) receptor. J. Biol. Chem. 277, 3236–3241.

Ehlers, A., Clark, D.M., 2000. A cognitive model of posttraumatic stress disorder. Behav. Res. Ther. 38, 319–345.

Esmaeili, A., Lynch, J.W., Sah, P., 2009. GABA$_A$ receptors containing gamma1 subunits contribute to inhibitory transmission in the central amygdala. J. Neurophysiol. 101, 341–349.

Eysenck, M.W., 1985. Anxiety and cognitive-task performance. Pers. Indiv. Dif. 6, 579–586.

File, S.E., Fluck, E., Joyce, E.M., 1999. Conditions under which lorazepam can facilitate retrieval. J. Clin. Psychopharmacol. 19, 349–353.

Filippov, A.K., Couve, A., Pangalos, M.N., Walsh, F.S., Brown, D.A., Moss, S.J., 2000. Heteromeric assembly of GABA(B)R1 and GABA (B)R2 receptor subunits inhibits Ca(2+) current in sympathetic neurons. J. Neurosci. 20, 2867–2874.

Fillmore, M.T., Kelly, T.H., Rush, C.R., Hays, L., 2001. Retrograde facilitation of memory by triazolam: effects on automatic processes. Psychopharmacology 158, 314–321.

Florey, E., 1991. GABA: history and perspectives. Can. J. Physiol. Pharmacol. 69, 1049–1056. (Abstract).

Foa, E.B., Keane, M.T., Friedman, M.J., 2000. Guidelines for the treatment of PTSD. J. Trauma. Stress 13, 539–588.

Fritschy, J.M., Benke, D., Mertens, S., Oertel, W.H., Bachi, T., Möhler, H., 1992. 5 Subtypes of type-A gamma-aminobutyric-acid receptors identified in neurons by double and triple immunofluoresence staining with subunit specific antibodies. Proc. Natl. Acad. Sci. USA 89, 6726–6730.

Galvez, T., Duthey, B., Kniazeff, J., Blahos, J., Rovelli, G., Bettler, B., et al., 2001. Allosteric interactions between GB1 and GB2 subunits are required for optimal GABA (B) receptor function. EMBO J. 20, 2152–2159.

Ghoneim, M.M., Mewaldt, S.P., 1975. Effects of diazepam and scopolamine on storage, retrieval and organizational processes in memory. Psychopharmacologia 44, 257–262.

Goudet, C., Magnaghi, V., Landry, M., Nagy, F., Gereau IV, R.W., Pin, J.P., 2009. Metabotropic receptors for glutamate and GABA in pain. Brain Res. Rev. 60, 43–56.

Hackmann, A., Holmes, E.A., 2004. Reflecting on imagery: a clinical perspective and overview of the special issue of Memory on mental imagery and memory in psychopathology. Memory 12, 389–402.

Hackmann, A., Clark, D.M., McManus, F., 2000. Recurrent images and early memories in social phobia. Behav. Res. Ther. 38, 601–610.

Haefely, W.E., 1989. Pharmacology of the benzodiazepine receptor. Eur. Arch. Psychiatr. Neurol. Sci. 238, 294–301.

Harris, J.A., Westbrook, R.F., 1998. Evidence that GABA transmission mediates context-specific extinction of learned fear. Psychopharmacology 140, 105–115.

Hassel, B., Dingledine, R., 2006. Glutamate. In: Siegel, G.J., Arganoff, B.W., Albers, R.W., Fisher, S.K., Uhler, M.D. (Eds.), Basic Neurochemistry: Molecular, Cellular, and Medical Aspects. Academic Press, New York, NY, pp. 267–290.

Hatfield, T., Spanis, C., McGaugh, J.L., 1999. Response of amygdalar norepinephrine to footshock and GABAergic drugs using in vivo microdialysis and HPLC. Brain Res. 835, 340–345.

Havlickova, M., Prezeau, L., Duthey, B., Bettler, B., Pin, J.P., Blahos, J., 2002. The intracellular loops of the GB2 subunit are crucial for G-protein coupling of the heteromeric gamma-aminobutyrate B receptor. Mol. Pharmacol. 62, 343–350.

Herzog, C.D., Stackman, R.W., Walsh, T.J., 1996. Intraseptal flumazenil enhances, while diazepam binding inhibitor impairs, performance in a working memory task. Neurobiol. Learn. Mem. 66, 341–352.

Hinrichs, J.V., Ghoneim, M.M., Mewaldt, S.P., 1984. Diazepam and memory: retrograde facilitation produced by interference reduction. Psychopharmacology 84, 158–162.

Houston, C.M., Smart, T.G., 2006. CaMK-II modulation of GABA(A) receptors expressed in HEK293, NG108-15 and rat cerebellar granule neurons. Eur. J. Neurosci. 24, 2504−2514.

Houston, C.M., Lee, H.H., Hosie, A.M., Moss, S.J., Smart, T.G., 2007. Identification of the sites for CaMK-II-dependent phosphorylation of GABA(A) receptors. J. Biol. Chem. 282, 17855−17865.

Huang, Y.W., Chen, Z., Hu, W.W., Zhang, L.S., Wu, W., Ying, L.Y., et al., 2003. Facilitating effect of histamine on spatial memory deficits induced by dizocilpine as evaluated by 8-arm radial maze in SD rats. Acta Pharmacol. Sin. 24, 1270−1276.

Huang, Y.W., Hu, W.W., Chen, Z., Zhang, L.S., Shen, H.Q., Timmerman, H., et al., 2004. Effect of the histamine H3-antagonist clobenpropit on spatial memory deficits induced by MK-801 as evaluated by radial maze in Sprague-Dawley rats. Behav. Brain Res. 151, 287−293.

Johnston, G.A., 1996. GABA$_A$ receptor pharmacology. Pharmacol. Ther. 69, 173−198.

Jorgensen, E.M., 2005. GABA. WormBook 31, 1−13.

Jurd, R., Moss, S.J., 2010. Impaired GABA$_A$ receptor endocytosis and its correlation to spatial memory deficits. Commun. Integr. Biol. 2, 176−178.

Katoh-Semba, R., Takeuchi, I.K., Inaguma, Y., Ichisaka, S., Hata, Y., Tsumoto, T., 2001. Induction of brain-derived neurotrophic factor by convulsant drugs in the rat brain: involvement of region-specific voltage-dependent calcium channels. J. Neurochem. 77, 71−83.

Kim, D.H., Kim, J.M., Park, S.J., Cai, M., Liu, X., Lee, S., et al., 2012. GABA$_A$ receptor blockade enhances memory consolidation by increasing hippocampal BDNF levels. Neuropsychopharmacology 37, 422−433.

Kniazeff, J., Galvez, T., Labesse, G., Pin, J.P., 2002. No ligand binding in the GB2 subunit of the GABA(B) receptor is required for activation and allosteric interaction between the subunits. J. Neurosci. 22, 7352−7361.

Korpi, E.R., Grunder, G., Luddens, H., 2002. Drug interactions at GABA(A) receptors. Prog. Neurobiol. 67, 113−159.

Lister, R.G., 1985. The amnesic action of benzodiazepines in man. Neurosci. Biobehav. Rev. 9, 87−94.

Lu, Y.M., Mansuy, I.M., Kandel, E.R., Roder, J., 2000. Calcineurin-mediated LTD of GABAergic inhibition underlies the increased excitability of CA1 neurons associated with LTP. Neuron. 26, 197−205.

Luft, T., Pereira, G.S., Cammarota, M., Izquierdo, I., 2004. Different time course for the memory facilitating effect of bicuculline in hippocampus, entorhinal cortex, and posterior parietal cortex of rats. Neurobiol. Learn. Mem. 82, 52−56.

Makkar, S.R., Zhang, S.Q., Cranney, J., 2010. Behavioral and neural analysis of GABA in the acquisition, consolidation, reconsolidation, and extinction of fear memory. Neuropsychopharmacology 35, 1625−1652.

Malenka, R.C., 1994. Synaptic plasticity in the hippocampus: LTP and LTD. Cell 78, 535−538.

Margeta-Mitrovic, M., Jan, Y.N., Jan, L.Y., 2001a. Function of GB1 and GB2 subunits in G protein coupling of GABA(B) receptors. Proc. Natl. Acad. Sci. USA 98, 14649−14654.

Margeta-Mitrovic, M., Jan, Y.N., Jan, L.Y., 2001b. Ligand-induced signal transduction within heterodimeric GABA(B) receptor. Proc. Natl. Acad. Sci. USA 98, 14643−14648.

McGaugh, J.L., Roozendaal, B., 2009. Drug enhancement of memory consolidation: historical perspective and neurobiological implications. Psychopharmacology (Berl.) 202, 3−14.

McNamara, R.K., Skelton, R.W., 1991. Diazepam impairs acquisition but not performance in the Morris water maze. Pharmacol. Biochem. Behav. 38, 651–658.

Mehta, A.K., Ticku, M.K., 1999. An update on GABA$_A$ receptors. Brain Res. Rev. 29, 196–217.

Metsis, M., Timmusk, T., Arenas, E., Persson, H., 1993. Differential usage of multiple brain-derived neurotrophic factor promoters in the rat brain following neuronal activation. Proc. Natl. Acad Sci. USA 90, 8802–8806.

Miller, M.M., Atemus, M., Debiec, J., LeDoux, J.E., 2004. Propranolol impairs reconsolidation of conditioned fear in humans. Soc. Neurosci. Abstracts 208, 2.

Möhler, H., Fritschy, J.M., Rudolph, U., 2002. A new benzodiazepine pharmacology. J. Pharmacol. Exp. Ther. 300, 2–8.

Möhler, H., Fritschy, J.M., Crestani, F., Hensch, T., Rudolph, U., 2004. Specific GABA(A) circuits in brain. Pharmacology 14, 363–367.

Moser, M.B., Moser, E.I., 1998. Distributed encoding and retrieval of spatial memory in the hippocampus. J. Neurosci. 18, 7535–7542.

Obradovic, D., Savic, M., Andjelkovic, D., Ugresic, N., Bokonjic, D., 2004. The influence of midazolam on active avoidance retrieval and acquisition rate in rats. Pharmacol. Biochem. Behav. 77, 77–83.

Olsen, R.W., Betz, H., 2006. GABA and glycine. In: Siegel, G.J., Arganoff, B.W., Albers, R.W., Fisher, S.K., Uhler, M.D. (Eds.), Basic Neurochemistry: Molecular, Cellular, and Medical Aspects. Academic Press, New York, NY, pp. 291–301.

Olsen, R.W., Sieghart, W., 2008. International union of pharmacology. LXX. Subtypes of γ-aminobutyric acid A receptors: classification on the basis of subunit composition, pharmacology, and function. Update Pharmacol. Rev. 60, 243–260.

Overington, J.P., Al-Lazikani, B., Hopkins, A.L., 2006. How many drug targets are there? Nat. Rev. Drug Discov. 5, 993–996.

Pain, L., Launoy, A., Fouquet, N., Oberling, P., 2002. Mechanisms of action of midazolam on expression of contextual fear in rats. Br. J. Anaesth. 89, 614–621.

Pin, J.P., Kniazeff, J., Binet, V., Liu, J., Maurel, D., Galvez, T., et al., 2004a. Activation mechanism of the heterodimeric GABA(B) receptor. Biochem. Pharmacol. 68, 1565–1572.

Pin, J.P., Kniazeff, J., Goudet, C., Bessis, A.S., Liu, J., Galvez, T., et al., 2004b. The activation mechanism of class-C G-protein coupled receptors. Biol. Cell 96, 335–342.

Pirker, S., Schwarzer, C., Wieselthaler, A., Sieghart, W., Sperk, G., 2000. GABA$_A$ receptors: immunocytochemical distribution of 13 subunits in the adult brain. Neuroscience 101, 815–850.

Rachman, S., 1991. Neo-conditioning and the classical theory of fear acquisition. Clin. Psychol. Rev. 11, 155–173.

Robello, M., Amico, C., Cupello, A., 1997. A dual mechanism for impairment of GABAA receptor activity by NMDA receptor activation in rat cerebellum granule cells. Eur. Biophys. J. 25, 181–187.

Roberts, E., 2000. Adventures with GABA: fifty years on. In: Martin, D.L., Olsen, R.W. (Eds.), GABA in the Nervous System: The View at Fifty Years. Lippincott Williams & Wilkins, Philadelphia, PA, pp. 1–24.

Rowley, N.M., Madsen, K.K., Schousboe, A, Steve White, H., 2012. Glutamate and GABA synthesis, release, transport and metabolism as targets for seizure control. Neurochem. Int. 61, 546–558.

Rudolph, U., Möhler, H., 2004. Analysis of GABA$_A$ receptor function and dissection of the pharmacology of benzodiazepines and general anesthetics through mouse genetics. Annu. Rev. Pharmacol. Toxicol. 44, 475–498.

Rudolph, U., Crestani, F., Möhler, H., 2001. GABA$_A$ receptor subtypes: dissecting their pharmacological functions. Trends Pharmacol. Sci. 22, 188−194.

Saito, S., Okada, A., Ouwa, T., Kato, A., Akagi, M., Kamei, C., 2010. Interaction between hippocampal g-aminobutyric acid A and N-methyl-D-aspartate receptors in the retention of spatial working memory in rats. Biol. Pharm. Bull. 33, 439−443.

Sanger, D., Willner, P., Bergman, J., 2003. Applications of behavioral pharmacology in drug discovery. Behav. Pharmacol. 14, 363−367.

Savić, M.M., Obradovic, D., Ugresic, N., Bokonjic, D., 2003. The influence of diazepam on atropine reversal of behavioral impairment in dichlorvos-treated rats. Pharmacol. Toxicol. 93, 211−218.

Savić, M.M., Obradovic, D.I., Ugresic, N.D., Cook, J.M., Yin, W., Bokonjic, D.R., 2005a. Bidirectional effects of benzodiazepine binding site ligands in the passive avoidance task: differential antagonism by flumazenil and 13-CCt. Behav. Brain Res. 158, 293−300.

Savić, M.M., Obradović, D.I., Ugresić, N.D., Cook, J.M., Sarma, P.V., Bokonjić, D.R., 2005b. Bidirectional effects of benzodiazepine binding site ligands on active avoidance acquisition and retention: differential antagonism by flumazenil and 13-CCt. Psychopharmacology 180, 455−465.

Savić, M.M., Obradović, D.I., Ugresić, N.D., Bokonjić, D.R., 2005c. Memory effects of benzodiazepines: memory stages and types versus binding-site subtypes. Neural Plast. 12, 289−298.

Schousboe, A., Waagepetersen, H.S., 2008. GABA neurotransmission: an overview. In: Lajtha, A. (Ed.), Handbook of Neurochemistry and Molecular Neurobiology. Springer, New York, NY, pp. 213−226.

Sieghart, W., Sperk, G., 2002. Subunit composition, distribution, and function of GABA$_A$ receptor subtypes. Curr. Top. Med. Chem. 2, 795−816.

Sigel, E., Steinmann, M.E., 2012. Structure, function, and modulation of GABA$_A$ receptors published. J. Biol. Chem. 287, 40224−40231.

Sramek, J.J., Zarotsky, V., Cutler, N.R., 2002. Generalised anxiety disorder: treatment options. Drugs 62, 1635−1648.

Stelzer, A., Shi, H., 1994. Impairment of GABA$_A$ receptor function by N-methyl-D-aspartate-mediated calcium influx in isolated CA1 pyramidal cells. Neuroscience. 62, 813−828.

Street, L.J., Sternfeld, F., Jelley, R.A., Reeve, A.J., Carling, R.W., Moore, K.W., et al., 2004. Synthesis and biological evaluation of 3-heterocyclyl-7,8,9,10-tetrahydro-(7,10-ethano)-1,2,4-triazolo [3,4-a]phthalazines and analogues as subtype-selective inverse agonists for the GABA(A)alpha5 benzodiazepine binding site. J. Med. Chem. 47, 3642−3657.

Tretter, V., Revilla-Sanchez, R., Houston, C., Terunuma, M., Havekes, R., Florian, C., et al., 2009. Deficits in spatial memory correlate with modified γ-aminobutyric acid type a receptor tyrosine phosphorylation in the hippocampus. Proc. Natl. Acad. Sci. USA 106, 20039−20044.

Venault, P., Chapouthier, G., de Carvalho, L.P., Simiand, J., Morre, M., Dodd, R.H., et al., 1986. Benzodiazepine impairs and beta-carboline enhances performance in learning and memory tasks. Nature 321, 864−866.

Vlachou, S., Markou, A., 2010. GABA$_B$ receptors in reward processes. Adv. Pharmacol. 58, 315−371.

Walls, A.B., Nilsen, L.H., Eyjolfsson, E.M., Vestergaard, H.T., Hansen, S.L., Schousboe, A., et al., 2010. GAD65 is essential for synthesis of GABA destined for tonic inhibition regulating epileptiform activity. J. Neurochem. 115, 1398−1408.

Walls, A.B., Eyjolfsson, E.M., Smeland, O.B., Nilsen, L.H., Schousboe, I., Schousboe, A., et al., 2011. Knockout of GAD65 has major impact on synaptic GABA synthesized from astrocyte-derived glutamine. J. Cereb. Blood Flow Metab. 31, 494−503.

Yerkes, R.M., Dodson, J.D., 1908. The relation of strength of stimulus to rapidity of habit-formation. J. Comp. Neurol. Psychol. 18, 459−482.

Yoshihara, T., Ichitani, Y., 2004. Hippocampal N-methyl-D-aspartate receptor-mediated encoding and retrieval processes in spatial working memory: delay-interposed radial maze performance in rats. Neuroscience 129, 1−10.

Zhang, S., Cranney, J., 2008. The role of GABA and pre-existing anxiety in the reconsolidation of conditioned fear. Behav. Neurosci. 122, 1295−1305.

4 Involvement of Glutamate in Learning and Memory

Antonella Gasbarri and Assunta Pompili

Department of Applied Clinical and Biotechnological Sciences, University of L'Aquila, Italy

Introduction*

The two most important neurotransmitters in the central nervous system (CNS) are glutamate and γ-aminobutyric acid (GABA), controlling excitatory and inhibitory neurotransmission, respectively (Hassel and Dingledine, 2006; Olsen and Betz, 2006; Rowley et al., 2012; Schousboe and Waagepetersen, 2008). Since its discovery in the early 1950s (Hayashi, 1954), glutamate has been implicated in many fundamental functions, including neuronal plasticity, neurotoxicity, development, learning and memory (Riedel et al., 2003). Synaptic transmission via glutamate receptors provides the excitatory drive for the neuronal pathways connecting the main cerebral regions (Roberts et al., 1981). Both ion channel coupled (ionotropic) and second messenger coupled (metabotropic) receptors are differentially distributed on pre- and postsynaptic sites, in order to participate to functions, such as neuronal communication and signal processing, whose importance is very evident, considering that they determine several cognitive functions, including learning and memory formation (Storm-Mathisen et al., 1983, 1995). The potential role of the different glutamate-receptor subtypes in long-term potentiation (LTP) and long-term depression (LTD) has extensively discussed (Bear and Abraham, 1996; Bliss and Collingridge, 1993; Riedel et al., 1996). Significant insight into the pharmacological properties of glutamate receptors and, therefore, valuable information about actions of specific agonists and antagonists were obtained by electrophysiological studies. Progresses in molecular and biochemical techniques have also given an important contribution to the detailed characterization of glutamate-receptor composition, function, and physiology.

Glutamate neurotransmission is also involved in hippocampal synaptic plasticity (Tamminga et al., 2012). It is very well known that hippocampal formation (HF) and the surrounding medial temporal lobe (MTL) cortex depend on glutamate signaling to a greater extent than other neocortical tissue, a feature that underlies its learning and memory functions (Amaral and Witter, 1989). Moreover, HF is one of the cerebral areas whose functions are altered in schizophrenia, suggesting the

Identification of Neural Markers Accompanying Memory. DOI: http://dx.doi.org/10.1016/B978-0-12-408139-0.00004-3

possible relevance of glutamate transmission in this structure to psychosis patho-physiology (Tamminga et al., 2010).

Considerable advances have also been reported for the role of glutamate recep-tors in many other diseases, including dissociative thought disorder, or various other forms of dementia such as Alzheimer's disease (Ellison, 1995; Pellicciari and Costantino, 1999).

Glutamate Receptors

Glutamate can bind to ionotropic (iGluRs) and metabotropic (mGluRs) receptors.

iGluRs mediate fast transmission and include three main classes: α-amino-3-hydroxy-5-methyl-4-isoxazolepropionic acid (AMPA) receptors, kainate, and N-methyl-D aspartate (NMDA) receptors, named after agonists that activate one particular class. Taking into account that only few pharmacological agents distin-guish between AMPA and kainate receptors (Wong et al., 1994), they are often referred to as "non-NMDA" receptors. iGluRs are tetra- or pentameric ion chan-nels, constituted of class-specific subunits, which can form homo- or heteromers. Each subunit has an extracellular N-terminus, while the carboxy terminus is expressed intracellularly.

mGluRs mediate transmission mechanisms, which are obviously slower and more modulatory in nature. So far, eight subtypes grouped into three classes have been characterized. mGluRs are assumed to have seven transmembrane regions similar to other G protein coupled receptors, while amino acid sequence similarities exist only with $GABA_B$ receptors.

Glutamatergic transmission and glutamate-receptor functions in learning and memory can be understood considering the composition of a given receptor, as well as the region- and even subcellular-specific distribution of GluRs. Beyond the morphological determinants, there are other ways of modifying receptor function, on the basis of extrinsic and intrinsic factors, such as application of drugs/toxins. Therefore, the situation is complex and many doubts still exist.

Ionotropic Receptors of Glutamate

AMPA Receptors

AMPA receptors (AMPARs) are heterotetrameric complexes composed of subunits GluA1−4 (Hollmann and Heinemann,1994), each of them conferring specific prop-erties on the active receptor, including trafficking motifs, which is related to synap-tic plasticity (Shepherd and Huganir, 2007). These glutamate receptors are responsible for the fast, immediate postsynaptic response to glutamate release. Therefore, the receptor−channel complex, predominantly permeable for sodium and potassium ions, activates fast and shows rapid desensitization.

Recent research has evidenced that AMPA plays a relevant role in long-term forms of synaptic plasticity such as LTP and LTD (Man, 2011).

The ionotropic AMPARs have usually the characteristic to be impermeable to calcium, which depends on the presence of GluA2 subunit. However, GluA2-lacking AMPARs are permeable to calcium and have recently been shown to play a unique role in synaptic activity.

Experiences and synaptic activity affect AMPARs (Shepherd, 2012). It is very well known that experiences shape and mold brain during the course of life, modifying nervous circuits through numerous cellular and molecular processes. Simplistically, neuronal circuits change by modifications in neurotransmitter release or neurotransmitter detection at synapses. Studies focused on the role of calcium permeable AMPA (CP-AMPA) receptors in experience-dependent and synaptic plasticity have suggested that CP-AMPARs play a prominent role in maintaining circuits in a labile state, where further plasticity can occur, thus promoting metaplasticity (Clem and Huganir, 2010; Herry et al., 2010; Shepherd, 2012). This "plasticity-ready" state can keep memory traces labile, but only in a time-delimited way. Indeed, long-term expression of CP-AMPARs has been related to neuronal pathologies, like ischemia, and therefore long-term expression of CP-AMPARs may have a detrimental effect on neuronal function (Liu and Zukin, 2007).

At the mechanistic level, very little is known concerning how the expression of CP-AMPARs influences neuronal properties and many questions remain unanswered, such as if specific second messenger systems exist, controlled through calcium flux via CP-AMPARs, or what are the trafficking rules governing activity-dependent CP-AMPAR expression, or in which way CP-AMPARs are specifically removed and replaced by GluA2-containing AMPARs. These questions are important not only to understand basic brain function, but also because CP-AMPARs may be an exciting new target for treatments of drug addiction and memory disorders.

Moreover, the abnormal expression of CP-AMPARs has been involved in drug addiction and memory deficits; therefore, these receptors could represent an exciting new target for therapeutics in drug addiction and memory disorders.

NMDA Receptors

While AMPARs have a more general importance for synaptic transmission per se, NMDA receptors (NMDARs) are considered the classic learning and memory receptors.

NMDARs are ubiquitously expressed throughout the CNS and are involved in many functions of neural circuits in the brain and spinal cord. They have an essential role in cognitive processes and they represent a prime target for cognitive enhancement (Collingridge et al., 2013).

Approximately 30 years ago, it was reported that the transient activation of NMDARs is the trigger for the induction of LTP at synapses existing between pyramidal neurons of CA1 and CA3 fields of HF (Collingridge et al., 1983).

Few years later, it was shown that NMDARs are also required for HF-dependent types of learning and memory (Morris et al., 1986). These findings have led to several studies on the role of NMDARs in synaptic plasticity, learning and memory and have considered the NMDARs centrally involved in cognitive processes. Since NMDARs are required for these mechanisms, the simple notion is that boosting NMDA function should enhance cognition and, indeed, there is evidence that this may be correct under certain circumstances (Tang et al., 1999).

Drugs stimulating NMDARs have been investigated in both animal models and clinical studies (Roesler and Schröder, 2011). Animal studies showed that NMDARs play a crucial role in various types of learning, such as eyeblink conditioning (Thompson and Disterhoft, 1997), reversal learning (Harder et al., 1998), Pavlovian fear conditioning (Xu et al., 2001), olfactory memory (Maleszka et al., 2000; Si et al., 2004), passive-avoidance learning (Danysz et al., 1988), spatial learning (Morris et al., 1986; Shimizu et al., 2000; Tsien et al., 1996), working and reference memory (Levin et al., 1998; May-Simera and Levin, 2003), and place preference (Swain et al., 2004). The activation of NMDARs is required for LTP in the HF, amygdala, and medial septum (Izquierdo, 1994; Rockstroh et al., 1996; Scatton et al., 1991). This mechanism is related to memory formation; the involvement of the glutamate-receptor system and LTP is strongly linked to new learning and memory in animal models (Lozano et al., 2001; Scheetz and Constantine-Paton, 1994; Tang et al., 1999; 2001; Wong et al., 2002). Both lesion studies and pharmacological manipulations in experimental animals suggest that the NMDAR may have a relevant role in the induction of memory formation but not in the conservation of memories (Constantine-Paton, 1994; Izquierdo, 1991; Izquierdo and Medina, 1993; Liang et al., 1993; Quartermain et al., 1994; Rickard et al., 1994). Indeed, it has been shown that blocking the NMDAR after learning does not affect memory performance in humans, whereas blockade of receptors before learning resulted in memory impairment (Hadj Tahar et al., 2004; Oye et al., 1992; Rowland et al., 2005).

Studies on transgenic and mutant mice has provided further evidence in support of the involvement of the NMDA in cognitive functions. Mutant mice lacking the NMDAR subunit NR2A have shown reduced hippocampal LTP and spatial learning (Sakimura et al., 1995). Also, transgenic mice lacking NMDARs in the CA1 region of the HF show both defective LTP and severe deficits in both spatial and non-spatial learning (Shimizu et al., 2000; Tsien et al., 1996). On the other hand, genetic enhancement of NMDA function results in better learning and memory. The role of the NMDAR in hippocampal LTP and in learning and memory was confirmed by studies showing larger hippocampal LTP and better learning and memory in NR2B transgenic mice, in which the NMDAR function is enhanced via the NR2B subunit transgene in forebrain neurons (Tang et al., 1999, 2001; Wong et al., 2002). However, in the novel-object recognition task, enriched NR2B transgenic mice exhibited much longer recognition memory (up to 1 week) compared with that of naïve NR2B transgenic mice (up to 3 days). Taken together, these findings confirm and support the important role of the NMDAR in memory processes (Rezvani, 2006).

The NMDAR system in the brain has also been implicated in learning and memory in humans, as it is evidenced by the results of several studies, using different

NMDA antagonists. NMDARs are involved in new memory formation in humans, while blockade of NMDARs after a task has been learned has no effect on memory in humans (Hadj Tahar et al., 2004; Oye et al., 1992; Rowland et al., 2005). Moreover, hypofunctionality of the glutamatergic system in the brain and, in particular, the hypofunction of the subpopulation of corticolimbic NMDARs have been implicated in the pathophysiology of schizophrenia (Coyle, 1996; Olney and Farber, 1995; Tsai and Coyle, 2002). The NMDAR antagonists ketamine and phencyclidine have been shown to evoke a range of symptoms and cognitive deficits similar to those observed in schizophrenia (Honey et al., 2005; Krystal et al., 1994).

Metabotropic Receptors of Glutamate

In the mid-1980s, it was reported that glutamate, besides its fast action on iGluR, also activates G protein coupled receptors linked to several intracellular second messenger cascades.

Since cloning of the first mGluR subtype, realized in the early1990s (Masu et al., 1991), 8 mGluR have been identified and divided into 3 groups: group I, including mGluR1 and mGluR5, group II which includes mGluR2 and mGluR3, and finally group III comprising mGluR4, mGluR6, mGluR7, and mGluR8 (Menard and Quirion, 2012; Nicoletti et al., 2011).

Differently by group I, group II, and group III mGluRs all are negatively linked to adenylate cyclase reducing the intracellular level of cyclic AMP. Group III mGluRs have also been identified as glutamate autoreceptors, then regulating release (Riedel et al., 2003).

The distribution of mGluRs is very heterogeneous throughout the brain (Shigemoto and Mizuno, 2000). Both in situ hybridization and immunolabeling showed distinct, but partly overlapping, distribution profiles. According to the anatomical distribution of glutamate in the brain, mGluRs may potentially participate in glutamatergic transmission on excitatory synapses in all behaviorally relevant brain structures, including HF, cortex, striatum, amygdala, and cerebellum (Moroni et al., 1998). More interestingly, the subcellular distribution of mGluRs shows striking differences. Group I mGluRs are localized postsynaptically on dendritic spines and are concentrated at perisynaptic sites, as revealed by electron microscopic immunogold labeling (Baude et al., 1993; Lujan et al., 1996). However, data also suggest a presynaptic modulation of transmitter release by group I mGluRs (Herrero et al., 1992).

Group II mGluRs are localized pre- and postsynaptically. Interestingly mGluR3 receptors are mainly found on glial cells. As suggested by their perisynaptic localization, group I and II mGluRs may not participate in normal synaptic transmission but become activated in situations of repeated stimulation of afferents and substantial glutamate accumulation in the synaptic cleft (Shigemoto et al., 1997). By contrast, group III mGluRs are mainly presynaptic at the axon terminal.

A growing number of compounds are available to investigate the pharmacology and functions of mGluR. The potential danger with these compounds is that intracerebral administration of mGluR agonists readily induces neurotoxicity (Saccan and Schoepp, 1992) so that doses need a very careful control if acute or chronic microinfusions are performed.

In parallel to the group-specific mGluR agonists, much effort was made to develop group and subtype-specific antagonists, often with the view of a possible clinical application in prevention of neurotoxicity.

So far, very little behavioral research has utilized group II and III antagonists. The main interest has focused on group I mGluR function in learning and memory, even though a role for group III mGluRs in behavioral flexibility and long-term memory storage has also been recognized (Gerlai et al., 1998).

The mGlu5 receptor is involved in learning and memory, contextual fear conditioning, inhibitory avoidance, fear potentiated startle, including spatial learning, and conditioned taste aversion (Simonyi et al., 2010). In the past years, important progress in behavioral pharmacology of mGlu5 receptor has been made. Findings suggest a task-specific involvement of mGlu5 receptor in learning and memory. The mGlu5 receptors have a central role in aversive learning and they are especially important for memory acquisition in avoidance and fear learning tasks. A role of the mGlu5 receptor in the consolidation and expression of fear memories has also been suggested. However, whether and how the mGlu5 receptor influences other memory phases, such as extinction and/or reconsolidation, has still to be determined.

To conclude, the findings of many studies evidenced that different mGluR subtypes have an important role in various aspects of learning-related neural processes. In particular, data suggest little or no involvement of mGluRs in the actual acquisition response. Instead, a relevant role is observed in long-term memory formation and there is strong evidence for modulatory actions on consolidation of memory. This is not only limited to episodic-like memories, but also includes procedural learning or habit formation, and sensory gating. Even though the function of the different mGluR subtypes may change depending on brain structure and task, in general posttraining activation of group II and III mGluRs enhances, while block or activation of group I mGluRs prevents consolidation processes and thus memory formation (Riedel et al., 2003). In contrast to NMDAR, mGluRs also are important in recall of memories, even when long (31 days) retention intervals are applied.

Glutamate, Memory, and HF

It is well known that HF and the surrounding MTL cortex depend on glutamate signaling to a greater extent than other neocortical tissue, a feature that underlies its important role in learning and memory (Amaral and Witter, 1989; Tamminga et al., 2012). We have already mentioned the involvement of group I of mGluRs in learning and memory. In particular, this group is involved in long-term memory formation in HF. Hippocampal mGlu5 receptor have a unique role in memory processing as suggested by studies showing the modification of function and/or expression of mGlu5 receptor after fear conditioning, eyeblink conditioning and maze learning. In addition, mGlu5 receptors in the lateral amygdala also make an important contribution to fear conditioning and conditioned taste aversion.

Group II mGluRs are required for persistent LTD in HF but are not necessary for LTP in the hippocampal CA1 region, *in vivo*. So far, the role of these receptors in spatial learning, and in synaptic plasticity in the dentate gyrus of HF *in vivo*, has not yet been the subject of extensive analysis. The effects of group II mGluR antagonism on LTP and LTD in the adult rat, at medial perforant path−dentate gyrus synapses and on spatial learning in the eight-arm radial maze, was investigated (Altinbilek and Manahan-Vaughan, 2009). Daily application of the group II mGluR antagonist (2S)-α-ethylglutamic acid impaired long-term (reference) memory with effects becoming apparent 6 days after training and drug-treatment began. Short-term memory was unaffected throughout the 10-day study. Acute injection of (2S)-α-ethylglutamic acid did not affect either LTD or LTP in the HF dentate gyrus *in vivo*; however, after six daily applications, an impairment of LTD, but not LTP, was shown.

These findings are in agreement with studies showing that prolonged antagonism of group II mGluRs induces an impairment of LTD, which parallels the deficits of spatial memory deficits, provoked by group II mGluR antagonists. Moreover, these data confirm the importance of group II mGluRs for spatial memory formation and offer a further link between the LTD and the encoding of spatial information in the HF.

Even though the role of mGlu5 receptors has been investigated in HF and amygdala, further studies are necessary to elucidate their involvement in selective brain regions, in different stages of memory formation, using a variety of learning tasks. In recent years, new, selective, and potent antagonists, agonists, and positive allosteric modulators of mGlu5 receptors were discovered. The next few years will certainly allow the extensive characterization of these drugs in behavioral models, in order to better understand the distinctive role of mGlu5 receptor in cognitive processes.

Glutamate and Addiction

Addiction is a chronic, relapsing disorder, characterized by the long-term tendency of addicted individuals to relapse. A major factor that obstructs the attainment of abstinence is the persistence of maladaptive drug-associated memories, which can retain drug seeking and taking behavior and stimulate unconscious relapse of these habits. Therefore, addiction can be conceptualized as a disorder of aberrant learning of the formation of strong instrumental memories, connecting actions to drug seeking and taking outcomes that ultimately are expressed as persistent stimulus-response habits (Milton and Everitt, 2012).

It has become increasingly evident that the cerebral circuits, neurotransmission systems, and cellular and molecular substrates mediating drug addiction have considerable overlap with those underlying normal learning and memory mechanisms (Berke and Hyman, 2000; Dalley and Everitt, 2009; Hyman et al., 2006;

Kelley, 2004; Koob, 2009; Nestler, 2002; Robbins et al., 2008). For example, acute and/or subchronic passive administration of various drugs of abuse such as cocaine, amphetamines, opiates, or alcohol can induce or modulate LTP or LTD in brain structures, such as the HF, and various meso-limbic areas. These same regions also mediate contextual, episodic, and emotional memory as well as spatial, habit, and incentive learning. The lasting neuro-adaptations, produced by chronic drug exposure, can lead to the perseveration of drug-seeking behaviors (Davis and Gould, 2008; Kalivas and O'Brien, 2008; Kauer, 2004; Kauer and Malenka, 2007), hyper-salience of drug-associated stimuli, and a learned "imprint" of drug use in the brain (Boning, 2009).

Glutamate plays a relevant role in regulating drug self-administration and drug-seeking behavior, and the past decade has witnessed a growing interest in the role of group I metabotropic glutamate receptors in mediating these behaviors. Group I mGlu receptors have a role in normal and drug-induced synaptic plasticity, drug reward, reinforcement and relapse-like behaviors, and addiction-related cognitive processes such as behavioral inflexibility, extinction learning maladaptive learning and memory (Olive, 2010). Animal models of addiction have showed that antagonists of group I mGlu receptors, particularly the mGlu5 receptor, reduce self-administration of virtually all drugs of abuse. Since inhibitors of mGlu5 receptor function are now utilized in clinical trials for other medical conditions and seem to be well tolerated, a key question that remains unanswered is related to the alterations in cognition, produced by these compounds, resulting in reduced drug intake and drug-seeking behavior.

Finally, recent findings have showed that, in contrast to mGlu5 receptor antagonists, positive allosteric modulation of mGlu5 receptors actually enhances synaptic plasticity and improves various aspects of cognition, including behavioral flexibility, extinction of drug-seeking behavior, and spatial learning. Thus, while inhibition of group I mGlu receptor function may reduce drug reward, reinforcement, and relapse-related behaviors, positive allosteric modulation of the mGlu5 receptor subtype may actually increase cognition and potentially reverse some of the cognitive deficits associated with chronic drug use (Olive, 2010).

Glutamatergic System and Nervous Diseases

Hypofunctionality of the glutamatergic system in the brain and, in particular, of the subpopulation of corticolimbic NMDARs have been involved in the pathophysiology of schizophrenia (Coyle, 1996; Olney and Farber, 1995; Tsai and Coyle, 2002; Rezvani, 2006). In order to test this hypothesis, the noncompetitive NMDA antagonist ketamine has been widely used for pharmacological manipulation of the NMDAR. Ketamine exposure causes the blockade of NMDARs and, as a consequence, hypofunctionality of the glutamatergic system in the brain. The effects of ketamine on cognitive function in humans have been investigated by several researchers (Honey et al., 2005; Krystal et al., 1994; Lisman et al., 1998; Malhotra et al., 1996; Newcomer and Krystal, 2001; Oye et al., 1992; Rockstroh et al., 1996;

Schugens et al., 1997). Taken as a whole, their findings demonstrated that NMDAR blockade in humans impairs learning and memory (Honey et al., 2005; Morgan et al., 2004; Rockstroh et al., 1996). The results of a functional magnetic resonance imaging (fMRI) study showed that acute ketamine exposure altered the brain response to executive demands in a verbal working-memory task. These data suggest a task-specific effect of ketamine on working memory in healthy volunteers (Honey et al., 2004). In another fMRI study using a double-blind, placebo-controlled, randomized within-subjects design, it was demonstrated that ketamine exposure disrupted frontal and hippocampal contributions to encoding and retrieval of episodic memory (Honey et al., 2005). In a double-blind study, it was evidenced that the low-affinity NMDAR channel blocker amantadine, given orally to healthy young volunteers, failed to block motor learning consolidation in subjects that had already learned the task (Hadj Tahar et al., 2004). It has also been reported that ketamine impaired learning of spatial and verbal information of healthy humans but not retrieval of information learned before ketamine administration (Rowland et al., 2005). In another double-blind placebo-controlled design with healthy human volunteers, ketamine treatment produced a dose-dependent impairment to episodic and working memory (Morgan et al., 2004). Ketamine also impaired recognition memory and procedural learning.

The evidence that HF is one of the brain regions whose function is altered in schizophrenia suggest the potential relevance of glutamate transmission in this structure to psychosis pathophysiology (Tamminga et al., 2010). The involvement of altered glutamate signaling in schizophrenia and the clinical and pharmacological basis for it have been widely accepted (Tamminga, 1998; Tamminga et al., 2010; Joseph, 2006). The glutamatergic hypothesis of schizophrenia has been highly generative for models of molecular pathology underlying the symptoms of the disease. A new model of HF dysfunction, which could generate psychosis and can be tested on human and animal models, was recently proposed (Tamminga et al., 2012). This model is based on evidence of altered HF learning and memory and suggest that altered glutamate-dependent functions in HF subfields are involved in the impaired declarative memory of schizophrenic patients (Tamminga et al., 2012).

The complexity and multiplicity of NMDARs, its heteromeric structure with intracellular sites for modulation of function and its subtypes provide several opportunities for therapeutic interventions. Due to the widely accepted role of NMDARs in plasticity and memory, the right approach for cognitive enhancement is potentiation of NMDAR function. However, taking into account the complex roles of NMDARs in synaptic transmission and bidirectional synaptic plasticity, the normalization of function could be a better strategy (Collingridge et al., 2013). Putative compounds for consideration should have subtle effects; overstimulation of NMDARs will likely cause several unwanted consequences, such as exacerbation of pain, hyperexcitability, and neurodegeneration. Therefore, subunit selectivity and limited efficacy may represent desirable properties. New compounds with direct but allosteric and specific effects on the NMDAR subunits may offer the most successful approach (Collingridge et al., 2013).

Conclusions

Even though the role of glutamate and its receptors in the different stages of learning and memory has been extensively investigated, further studies are necessary to better clarify their distinctive roles. Moreover, the rapid advances of basic neuroscience in the next decade should provide more data on the complex mechanisms underlying schizophrenia and other nervous diseases. Efforts to develop new cognitive-enhancing drugs are likely to continue to include a major focus on glutamate and its receptors and their downstream signaling pathways.

References

Altinbilek, B., Manahan-Vaughan, D., 2009. A specific role for group II metabotropic glutamate receptors in hippocampal long-term depression and spatial memory. Neuroscience 158, 149−158.

Amaral, D.G., Witter, M.P., 1989. The three-dimensional organization of the hippocampal formation: a review of the anatomical data. Neuroscience 31, 571−591.

Baude, A., Nusser, Z., Roberts, J.D., Mulvihill, E., McIlhinney, R.A., Somogyi, P., 1993. The metabotropic glutamate receptor (mGluR1a) is concentrated at perisynaptic membrane of neuronal subpopulations as detected by immunogold reaction. Neuron 11, 771−787.

Bear, M.F., Abraham, W.C., 1996. Long-term depression in hippocampus. Ann. Rev. Neurosci. 19, 437−462.

Berke, J.D., Hyman, S.E., 2000. Addiction, dopamine and the molecular mechanisms of memory. Neuron 25, 515−532.

Bliss, T.V.P., Collingridge, G.L., 1993. A synaptic model of memory: long-term potentiation in the hippocampus. Nature 36, 31−39.

Boning, J., 2009. Addiction memory as a specific, individually learned memory imprint. Pharmacopsychiatry 42, S66−S68.

Clem, R.L., Huganir, R.L., 2010. Calcium-permeable AMPA receptor dynamics mediate fear memory erasure. Science 330, 1108−1112.

Collingridge, G.L., Kehl, S.J., McLennan, H., 1983. Excitatory amino acids in synaptic transmission in the Schaffer collateral-commissural pathway of the rat hippocampus. J. Physiol. 334, 33−46.

Collingridge, G.L., Volianskis, A., Bannister, N., France, G., Hanna, L., Mercier, M., et al., 2013. The NMDA receptor as a target for cognitive enhancement. Neuropharmacology 64, 13−26.

Constantine-Paton, M., 1994. Effects of NMDA receptor antagonists on the developing brain. Psychopharmacol. Bull 30, 561−565.

Coyle, J.T., 1996. The glutamatergic dysfunction hypothesis for schizophrenia. Harv. Rev. Psychiatry 3, 241−253.

Dalley, J.W., Everitt, B.J., 2009. Dopamine receptors in the learning, memory and drug reward circuitry. Semin. Cell Dev. Biol. 20, 403−410.

Danysz, W., Wroblewski, J.T., Costa, E., 1988. Learning impairment in rats by N-methyl-D-aspartate receptor antagonists. Neuropharmacology 27, 653−656.

Davis, J.A., Gould, T.J., 2008. Associative learning, the hippocampus, and nicotine addiction. Curr. Drug Abuse Rev. 1, 9–19.

Ellison, G., 1995. The N-methyl-D-aspartate antagonists phencyclidine, ketamine and dizocilpine as both behavioral and anatomical models of the dementias. Brain Res. Rev. 20, 250–267.

Gerlai, R., Roder, J.C., Hampson, D.R., 1998. Altered spatial learning and memory in mice lacking the mGluR4 subtype of metabotropic glutamate receptor. Behav. Neurosci. 112, 525–532.

Hadj Tahar, A., Blanchet, P.J., Doyon, J., 2004. Motor-learning impairment by amantadine in healthy volunteers. Neuropsychopharmacology. 29, 187–194.

Harder, J.A., Aboobaker, A.A., Hodgetts, T.C., Ridley, R.M., 1998. Learning impairments induced by glutamate blockade using dizocilpine (MK-801) in monkeys. Br. J. Pharmacol. 125, 1013–1018.

Hassel, B., Dingledine, R., 2006. Glutamate. In: Siegel, G.J., Arganoff, B.W., Albers, R.W., Fisher, S.K., Uhler, M.D. (Eds.), Basic Neurochemistry: Molecular, Cellular, and Medical Aspects. Academic Press, New York, NY, pp. 267–290.

Hayashi, T., 1954. Effects of sodium glutamate on the nervous system. Keio J. Med. 3, 192–193.

Herrero, I., Miras-Portugal, M.T., Sanchez-Prieto, J., 1992. Positive feedback of glutamate exocytosis by metabotropic glutamate receptor stimulation. Nature 360, 162–166.

Herry, C., Ferraguti, F., Singewald, N., Letzkus, J.J., Ehrlich, I., Luthi, A., 2010. Neuronal circuits of fear extinction. Eur. J. Neurosci. 31, 599–612.

Hollmann, M., Heinemann, S., 1994. Cloned glutamate receptors. Annu. Rev. Neurosci. 17, 31–108.

Honey, G.D., Honey, R.A., O'Loughlin, C., Sharar, S.R., Kumaran, D., Suckling, J., et al., 2005. Ketamine disrupts frontal and hippocampal contribution to encoding and retrieval of episodic memory: an fMRI study. Cereb. Cortex 15, 749–759.

Honey, R.A., Honey, G.D., O'Loughlin, C., Sharar, S.R., Kumaran, D., Bullmore, E.T., et al., 2004. Acute ketamine administration alters the brain responses to executive demands in a verbal working memory task: an fMRI study. Neuropsychopharmacology 29, 1203–1214.

Hyman, S.E., Malenka, R.C., Nestler, E.J., 2006. Neural mechanisms of addiction: the role of reward-related learning and memory. Annu. Rev. Neurosci. 29, 565–598.

Izquierdo, I., 1991. Role of NMDA receptors in memory. Trends Pharm. Sci. 12, 128–129.

Izquierdo, I., 1994. Pharmacological evidence for a role of long-term potentiation in memory. FASEB J. 8, 1139–1145.

Izquierdo, I., Medina, J.H., 1993. Role of the amygdala, hippocampus and entorhinal cortex in memory consolidation and expression. Braz. J. Med. Biol. Res. 26, 573–589.

Joseph, T.C., 2006. Glutamate and schizophrenia: beyond the dopamine hypothesis. Cell Mol. Neurobiol. V26, 363–382.

Kalivas, P.W., O'Brien, C., 2008. Drug addiction as a pathology of staged neuroplasticity. Neuropsychopharmacology 33, 166–180.

Kauer, J.A., 2004. Learning mechanisms in addiction: synaptic plasticity in the ventral tegmental area as a result of exposure to drugs of abuse. Annu. Rev. Physiol. 66, 447–475.

Kauer, J.A., Malenka, R.C., 2007. Synaptic plasticity and addiction. Nat. Rev. Neurosci. 8, 844–858.

Kelley, A.E., 2004. Memory and addiction: shared neural circuitry and molecular mechanisms. Neuron 44, 161–179.

Koob, G.F., 2009. Dynamics of neuronal circuits in addiction: reward, antireward, and emotional memory. Pharmacopsychiatry 42, S32–S41.

Krystal, J.H., Karper, L.P., Seibyl, J.P., Freeman, G.K., Delaney, R., Bremner, J.D., et al., 1994. Subanesthetic effects of the noncompetitive NMDA antagonist, ketamine, in humans: psychotomimetic, perceptual, cognitive, and neuroendocrine responses. Arch Gen. Psychiatry 51, 199–214.

Levin, E.D., Bettegowda, C., Weaver, T., Christopher, N.C., 1998. Nicotine–dizocilpine interactions and working and reference memory performance of rats in the radial arm maze. Pharmacol. Biochem. Behav. 61, 335–340.

Liang, K.C., Lin, M.H., Tyan, Y.M., 1993. Involvement of amygdala N-methyl-D-aspartate receptors in long-term retention of an inhibitory avoidance response in rats. Chin. J. Physiol. 36, 47–56.

Lisman, J.E., Fellous, J.M., Wang, X.J., 1998. A role for NMDA-receptor channels in working memory. Nat. Neurosci. 1, 273–275.

Liu, S.J., Zukin, R.S., 2007. Ca^{2+}-permeable AMPA receptor sinsynaptic plasticity and neuronal death. Trends Neurosci. 30, 126–134.

Lozano, V.C., Armengaud, C., Gauthier, M., 2001. Memory impairment induced by cholinergic antagonists injected into the mushroom bodies of the honeybee. J. Comp. Physiol. 187, 249–254.

Lujan, R., Nusser, Z., Roberts, J.D.B., Shigemoto, R., Somogyi, P., 1996. Perisynaptic location of metabotropic glutamate receptors mGluR1 and mGluR5 on dendrites and dendritic spines in the rat hippocampus. Eur. J. Neurosci. 8, 1488–1500.

Maleszka, R., Helliwell, P., Kucharski, R., 2000. Pharmacological interference with glutamate re-uptake impairs long-term memory in the honeybee, Apis mellifera. Behav. Brain Res. 115, 49–53.

Malhotra, A.K., Pinals, D.A., Weingartner, H., Sirocco, K., Missar, C.D., Pickar, D., et al., 1996. NMDA receptor function and human cognition: the effects of ketamine in healthy volunteers. Neuropsychopharmacology 14, 301–307.

Man, H.Y., 2011. GluA2-lacking, calcium-permeable AMPA receptors–inducers of plasticity? Curr. Opin. Neurobiol. 21, 291–298.

Masu, M., Tanabe, Y., Tsuchida, K., Shigemoto, R., Nakanishi, S., 1991. Sequence and expression of a metabotropic glutamate receptor. Nature 349, 760–765.

May-Simera, H., Levin, E.D., 2003. NMDA systems in the amygdala and piriform cortex and nicotinic effects on memory function. Brain Res. Cogn. Brain Res. 17, 475–483.

Ménard, C., Quirion, R., 2012. Group 1 metabotropic glutamate receptor function and its regulation of learning and memory in the aging brain. Front. Pharmacol. 3, 182.

Milton, A.L., Everitt, B.J., 2012. The persistence of maladaptive memory: addiction, drug memories and anti-relapse treatments. Neurosci. Biobehav. Rev. 36, 1119–1139.

Morgan, C.J., Mofeez, A., Brandner, B., Bromley, L., Curran, H.V., 2004. Acute effects of ketamine on memory systems and psychotic symptoms in healthy volunteers. Neuropsychopharmacology 29, 208–218.

Moroni, F., Nicoletti, F., Pellegrini-Gampietro, D.E., 1998. Metabotropic Glutamate Receptors in Brain Function. Portland Press, London.

Morris, R.G., Anderson, E., Lynch, G.S., Baudry, M., 1986. Selective impairment of learning and blockade of long-term potentiation by an N-methyl-D-aspartate receptor antagonist, AP5. Nature 319, 774–776.

Nestler, E.J., 2002. Common molecular and cellular substrates of addiction and memory. Neurobiol. Learn Mem. 78, 637–647.

Newcomer, J.W., Krystal, J.H., 2001. NMDA receptor regulation of memory and behavior in humans. Hippocampus 11, 529–542.

Nicoletti, F., Bockaert, J., Collingridge, G.L., Conn, P.J., Ferraguti, F., Schoepp, D.D., et al., 2011. Metabotropic glutamate receptors: from the workbench to the bedside. Neuropharmacology 60, 1017–1041.

Olive, M.F., 2010. Cognitive effects of Group I metabotropic glutamate receptor ligands in the context of drug addiction. Eur. J. Pharmacol. 639, 47–58.

Olney, J.W., Farber, N.B., 1995. NMDA antagonists as neurotherapeutic drugs, psychotogens, neurotoxins, and research tools for studying schizophrenia. Neuropsychopharmacology 13, 335–345.

Olsen, R.W., Betz, H., 2006. GABA and glycine. In: Siegel, G.J., Arganoff, B.W., Albers, R. W., Fisher, S.K., Uhler, M.D. (Eds.), Basic Neurochemistry: Molecular, Cellular, and Medical Aspects. Academic Press, New York, NY, pp. 291–301.

Oye, I., Paulsen, O., Maurset, A., 1992. Effects of ketamine on sensory perception: evidence for a role of N-methyl-D-aspartate receptors. J. Pharmacol. Exp. Ther. 260, 1209–1213.

Pellicciari, R., Costantino, G., 1999. Metabotropic G-protein coupled glutamate receptors as therapeutic targets. Curr. Opin. Chem. Biol. 3, 433–440.

Quartermain, D., Mower, J., Rafferty, M.F., Herting, R.L., Lanthorn, T.H., 1994. Acute but not chronic activation of the NMDA-coupled glycine receptor with D-cycloserine facilitates learning and retention. Eur. J. Pharmacol. 257, 7–12.

Rezvani, A.H., 2006. In: Levin, E.D., Buccafusco, J.J. (Eds.), Involvement of the NMDA System in Learning and Memory. Animal Models of Cognitive Impairment. CRC Press, Boca Raton, FL.

Rickard, N.S., Poot, A.C., Gibbs, M.E., Ng, K.T., 1994. Both non-NMDA and NMDA glutamate receptors are necessary for memory consolidation in the day-old chick. Behav. Neural Biol. 62, 33–40.

Riedel, G., Wetzel, W., Reymann, K.G., 1996. Comparing the role of metabotropic glutamate receptors in long-term potentiation and in learning and memory. Progr. Neuropsychopharmacol. Biol. Psychiatry 20, 761–789.

Riedel, G., Platt, B., Micheau, J., 2003. Glutamate receptor function in learning and memory. Behav. Brain Res. 140, 1–47.

Robbins, T.W., Ersche, K.D., Everitt, B.J., 2008. Drug addiction and the memory systems of the brain. Ann. NY Acad. Sci. 1141, 1–21.

Roberts, P.J., Storm-Mathisen, J., Johnstonm, G.A.R. (Eds.), 1981. Glutamate: Transmitter in the Central Nervous System. Wiley, Chichester, UK.

Rockstroh, S., Emre, M., Tarral, A., Pokorny, R., 1996. Effects of the novel NMDA-receptor antagonist SDZ EAA 494 on memory and attention in humans. Psychopharmacology 124, 261–266.

Roesler, R., Schröder, N., 2011. Cognitive enhancers: focus on modulatory signaling influencing memory consolidation. Pharmacol. Biochem. Behav. 99, 155–163.

Rowland, L.M., Astur, R.S., Jung, R.E., Bustillo, J.R., Lauriello, J., Yeo, R.A., 2005. Selective cognitive impairments associated with NMDA receptor blockade in humans. Neuropsychopharmacology 30, 633–639.

Rowley, N.M., Madsen, K.K., Schousboe, A, Steve White, H., 2012. Glutamate and GABA synthesis, release, transport and metabolism as targets for seizure control. Neurochem. Int. 61, 546–558.

Saccan, A.I., Schoepp, D.D., 1992. Activation of hippocampal metabotropic excitatory amino acid receptors leads to seizures and neuronal damage. Neurosci. Lett. 139, 77–82.

Sakimura, K., Kutsuwada, T., Ito, I., Manabe, T., Takayama, C., Kushiya, E., et al., 1995. Reduced hippocampal LTP and spatial learning in mice lacking NMDA receptor epsilon 1 subunit. Nature 373, 151–155.

Scatton, B., Carter, C., Benavides, J., 1991. NMDA receptor antagonists: treatment of brain ischemia. Drug News Persp. 4, 89–102.

Scheetz, A.J., Constantine-Paton, M., 1994. Modulation of NMDA receptor function: implications for vertebrate neural development. FASEB J. 8, 745–752.

Schousboe, A., Waagepetersen, H.S., 2008. GABA neurotransmission: an overview. In: Lajtha, A. (Ed.), Handbook of Neurochemistry and Molecular Neurobiology. Springer, New York, NY, pp. 213–226.

Schugens, M.M., Egerter, R., Daum, I., Schepelmann, K., Klockgether, T., Loschmann, P.A., 1997. The NMDA antagonist memantine impairs classical eyeblink conditioning in humans. Neurosci. Lett. 224, 57–60.

Shepherd, J.D., 2012. Memory, plasticity, and sleep—role for calcium permeable AMPA receptors? Front. Mol. Neurosci. 5, 49.

Shepherd, J.D., Huganir, R.L., 2007. The cell biology of synaptic plasticity: AMPA receptor trafficking. Annu. Rev. Cell. Dev. Biol. 23, 613–643.

Shigemoto, R., Mizuno, N., 2000. Metabotropic glutamate receptors immunocytochemical and in situ hybridisation analyses, Handbook of Chemical Neuroanatomy, Elsevier, vol. 18. pp. 63–98.

Shigemoto, R., Kinoshita, A., Wada, E., Nomura, S., Ohishi, H., Takada, M., et al., 1997. Differential presynaptic localization of metabotropic glutamate receptor subtypes in the rat hippocampus. J. Neurosci. 17, 7503–7522.

Shimizu, E., Tang, Y.P., Rampon, C., Tsien, J.Z., 2000. NMDA receptor-dependent synaptic reinforcement as a crucial process for memory consolidation. Science 290, 1170–1174.

Si, A., Helliwell, P., Maleszka, R., 2004. Effects of NMDA receptor antagonists on olfactory learning and memory in the honeybee (Apis mellifera). Pharmacol. Biochem. Behav. 77, 191–197.

Simonyi, A., Schachtman, T.R., Christoffersen, G.R.J., 2010. Metabotropic glutamate receptor subtype 5 antagonism in learning and memory. Eur. J. Pharmacol. 639, 17–25.

Storm-Mathisen, J., Leknes, A.K., Bore, A., Vaaland, J.L., Edminson, P., Haug, F.M.S., et al., 1983. First visualisation of glutamate and GABA in neurones by immunocytochemistry. Nature 301, 517–520.

Storm-Mathisen, J., Danbolt, N.C., Ottersen, O.P., 1995. Localisation of glutamate and its membrane transport proteins. In: Stone, T.W. (Ed.), CNS Neurotransmitters Neuromodulators. CRC Press, Boca Raton, FL, pp. 1–18.

Swain, H.A., Sigstad, C., Scalzo, F.M., 2004. Effects of dizocilpine (MK-801) on circling behavior, swimming activity, and place preference in zebrafish (Danio rerio). Neurotoxicol. Teratol. 26, 725–729.

Tamminga, C.A., 1998. Schizophrenia and glutamatergic transmission. Crit. Rev. Neurobiol. 12, 21–36.

Tamminga, C.A., Stan, A.D., Wagner, A.D., 2010. The hippocampal formation in schizophrenia. Am. J. Psychiatry 167, 1–16.

Tamminga, C.A., Southcott, S., Sacco, C., Wagner, A.D., Ghose, S., 2012. Glutamate dysfunction in hippocampus: relevance of dentate gyrus and CA3 signaling. Schizophrenia Bull 38, 927–935.

Tang, Y.P., Shimizu, E., Dube, G.R., Rampon, C., Kerchner, G.A., Zhuo, M., et al., 1999. Genetic enhancement of learning and memory in mice. Nature 401, 63–69.

Tang, Y.P., Wang, H., Feng, R., Kyin, M., Tsien, J.Z., 2001. Differential effects of enrichment on learning and memory function in NR2B transgenic mice. Neuropharmacology 41, 779−790.

Thompson, L.T., Disterhoft, J.F., 1997. N-methyl-D-aspartate receptors in associative eyeblink conditioning: both MK-801 and phencyclidine produce task- and dose-dependent impairments. J. Pharmacol. Exp. Ther. 28, 928−940.

Tsai, G., Coyle, J.T., 2002. Glutamatergic mechanisms in schizophrenia. Ann. Rev. Pharmacol. Toxicol. 42, 165−179.

Tsien, J.Z., Huerta, P.T., Tonegawa, S., 1996. The essential role of hippocampal CA1 NMDA receptor-dependent synaptic plasticity in spatial memory. Cell 87, 1327−1338.

Wong, L.A., Mayer, M.L., Jane, D.E., Watkins, J.C., 1994. Willardiines differentiate agonist-binding sites for kainate versus AMPA preferring glutamate receptors in DRG and hippocampal neurons. J. Neurosci. 14, 3881−3897.

Wong, R.W., Setou, M., Teng, J., Takei, Y., Hirokawa, N., 2002. Over expression of motor protein KIF17 enhances spatial and working memory in transgenic mice. Proc. Natl. Acad. Sci. USA 99, 14500−14505.

Xu, X., Russell, T., Bazner, J., Hamilton, J., 2001. NMDA receptor antagonist AP5 and nitric oxide synthase inhibitor 7-NI affect different phases of learning and memory in goldfish. Brain Res. 889, 274−277.

5 Dopamine and Memory

Ryan T. LaLumiere

Department of Psychology, University of Iowa, Iowa City, IA 52242

Through a series of studies in the 1950s and 1960s, the monoamine class of neuro-transmitters was discovered and the basic neurobiology underlying them was characterized (Carlsson et al., 1957, 1962; Carlsson, 1987). Anatomy studies noted that while the monoamine cells were relatively few in number in the brain, they had extensive projections throughout the brain, enabling them to modulate activity in widespread regions (Anden et al., 1964; Dahlstroem and Fuxe, 1964; Ungerstedt, 1971). As a result, studies have examined how these systems might have roles in general processes throughout the brain that require changes in activity across a number of different structures, including those involved in learning and memory. This chapter will specifically focus on how the dopamine system influences memory processes. The first part of the chapter will provide a description of the pharmacology of dopamine, including discussion of its receptors and known signaling mechanisms. The chapter will then discuss the anatomy of the dopamine system, particularly examining the two main dopamine projections: the mesocorticolimbic and the nigrostriatal systems. Following that, the chapter will proceed to examine several areas of memory research in which dopamine plays a critical role, including reward signaling, working memory, and long-term plasticity and memory consolidation.

Pharmacology of Dopamine

Dopamine, like other neurotransmitters, is stored in synaptic vesicles and is typically released in an activity-dependent manner. There are several dopamine receptor subtypes in the mammalian brain, identified as D1–D5, that are grouped into two general categories: D1- and D2-like receptors (Andersen et al., 1990; Niznik and Van Tol, 1992). The D1-like receptors include the D1 and D5 receptor subtypes and are generally considered excitatory, whereas the D2-like receptors include the D2, D3, and D4 receptor subtypes and are considered inhibitory. The D1-like receptors (hereafter referred to simply as "D1") are $G_{s/olf}$-coupled receptors that increase cAMP levels and activate protein kinase A (PKA). In contrast, the D2-like receptors (hereafter "D2") are $G_{i/o}$-coupled receptors that decrease cAMP levels. However, evidence suggests that some of the receptor subtypes may also function through different signaling pathways, depending on the cells expressing

Identification of Neural Markers Accompanying Memory. DOI: http://dx.doi.org/10.1016/B978-0-12-408139-0.00005-5

the receptor. For example, evidence suggests that activation of D1-like receptors stimulates phospholipase C activity independently of cAMP in renal cells (Felder et al., 1989). D2 receptors have also been found to regulate phospholipase C and intracellular calcium levels via the $G\beta\gamma$ subunit (Hernandez-Lopez et al., 2000). Moreover, evidence suggests that dopamine receptors directly interact with various ion channels, including subunits of the N-methyl-D-aspartate (NMDA) receptor, within different brain regions (Beaulieu and Gainetdinov, 2011). D1 receptors are typically found on the postsynaptic cell but can be located presynaptically, functioning to increase transmitter release, whereas D2 receptors are located both presynaptically, functioning to inhibit neurotransmitter release, and postsynaptically.

Dopamine is typically cleared from the synapse through two principal means. The dopamine transporter (DAT) acts via reuptake to remove synaptic dopamine. In addition, the extracellular enzyme monoamine oxidase breaks down synaptic dopamine. Intriguingly, while DAT is believed to be the larger contributor to dopamine clearance, the degree to which DAT is responsible for dopamine clearance appears to vary across the brain (Ciliax et al., 1995). For example, although the ventral and dorsal striatum rely heavily on DAT, the medial prefrontal cortex (PFC) also relies significantly on monamine oxidase for dopamine clearance (Wayment et al., 2001). Most likely, these regional differences in dopamine clearance alter the kinetics of synaptic dopamine and subsequent receptor activation, as dopamine may remain longer in the synaptic cleft in some regions and even act in a paracrine fashion. The functional importance of these regional differences remains largely unclear.

Anatomy of the Dopaminergic System

Dopamine-producing neurons are present within only a few cell groups in the central nervous system. The major groups are located in the A8, A9, and A10 cell groups within the midbrain and collectively form the ventral tegmental area (VTA)−substantia nigra (SN), the main focus of the present chapter (Ungerstedt, 1971; Fallon and Loughlin, 1995). In contrast, the A12 and A13 cell groups, which also produce dopamine, are found in the hypothalamus, whereas a third group, the A11 cells, is located in the subparafasicular thalamic nucleus (Ungerstedt, 1971; Takada, 1990). The SN is divided into two components, the pars compacta (SNc) and the pars reticulata (SNr). The SNc comprises the A9 dopamine neurons, whereas the SNr contains GABAergic cells that lie immediately ventral to the SNc as well as the dendrites of SNc neurons. The A9 dopamine cells extend caudally to form the A8 dopamine cells that make up the retrorubral field. In contrast, the A10 dopamine cells compose the VTA but, it should be noted, the VTA contains a significant number of nondopamine neurons. Although previous evidence suggested the VTA was primarily composed of dopamine and GABAergic cells, recent findings suggest a more complicated picture. Indeed, evidence indicates the presence of VTA and SNc cells expressing the vesicular glutamate transporter (VGluT), a marker of glutamatergic neurons, and physiological studies have found that cells

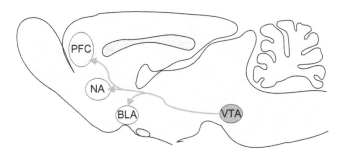

Figure 5.1 Sagittal section from a rat brain depicting part of the mesocorticolimbic
pathway. The VTA provides the dopaminergic innervation for most of the forebrain,
including the BLA, the NA, and the PFC. VTA, ventral tegmental area; BLA, basolateral
amygdala; NA, nucleus accumbens; PFC, prefrontal cortex.

originating in the VTA release glutamate at their efferent targets (Nair-Roberts
et al., 2008; Hnasko et al., 2010; Yamaguchi et al., 2011). Of particular interest, it
appears that a number of the VGluT-expressing neurons also express tyrosine
hydroxylase (Descarries et al., 2008; Li et al., 2012), suggesting the existence of
glutamate/dopamine neurons. These findings are relatively recent and the functional consequence of this anatomy is still being determined.

The VTA and SNc project throughout the forebrain and are responsible for most
of the dopaminergic innervation of the forebrain. The SNc provides a highly dense
innervation of the dorsal striatum (i.e., the caudate−putamen) with little dopamine
innervation of other structures (Gerfen et al., 1987; Fallon and Loughlin, 1995). In
contrast, the VTA projects to regions of the limbic system, including the hippocampus, the amygdala, and the bed nucleus of the stria terminalis, as well as to a variety of cortical regions, including the PFC, and thus, the entire VTA innervation of
the forebrain is known as the "mesocorticolimbic" system (Fallon and Loughlin,
1995; Gasbarri et al., 1997; Asan, 1998). Although the dorsal striatum receives its
dopamine input from the SNc, the ventral striatum, comprising the nucleus accumbens (NA) and olfactory tubercle, receives dopamine from the VTA projections,
suggesting the ventral striatum as a more "limbic system"-oriented structure.
Figure 5.1 provides a sagittal section from a rat brain showing the pathways from
the VTA to some of the important structures that will be discussed later in the
chapter. The VTA and SNc retain a significant topographic mapping with regard to
both inputs and outputs, as distinct groups of neurons within the structures receive
inputs from specific structures and project to distinct structures within the forebrain. Thus, for example, the dopamine innervation of the basolateral amygdala
(BLA) appears to originate from the lateral VTA as well as a small portion of the
adjacent medial SNc (Fallon et al., 1978).

The VTA−SNc receive inputs from a number of different structures throughout
the forebrain and brainstem. Inputs arise from the PFC, the amygdala, the subthala-
mic nucleus, and the brainstem nuclei containing serotonergic, cholinergic, and

noradrenergic inputs (Grillner and Mercuri, 2002). In addition, as part of the basal ganglia circuit, the VTA−SNc receive inputs from the striatum (ventral and dorsal) and pallidum (ventral and globus pallidus) (Fonnum et al., 1978; Bolam and Smith, 1990; Smith and Bolam, 1990). As described later, in the section on the basal ganglia, the striatal and pallidal inputs to the VTA−SNc follow a topographic gradient in which they innervate dopamine cells that provide reciprocal innervations of the same regions in the striatum and pallidum.

Within specific receptor subtype distributions, D1 receptors can be found at moderate-to-dense levels throughout the target regions of the mesocorticolimbic and nigrostrial pathways (Beaulieu and Gainetdinov, 2011). D5 receptors, in contrast, are expressed at considerably lower levels in cortical regions, the hippocampus, and the striatum (Missale et al., 1998; Beaulieu and Gainetdinov, 2011). D2 receptors are found at dense levels throughout the striatum and are also located throughout cortical and limbic targets as well as within the midbrain (VTA−SN), whereas D3 receptors appear to be limited to specific limbic regions including the NAshell and the midbrain dopamine neurons (Beaulieu and Gainetdinov, 2011). D4 receptors, while expressed at low levels in the striatum and pallidal structures, have higher expression levels in cortical and limbic targets, such as the amygdala and hippocampus (Beaulieu and Gainetdinov, 2011).

Basal Ganglia

The dopamine system is a critical component of the basal ganglia, a set of subcortical structures that govern movement, specific types of learning, and motivation. At the center of the basal ganglia is the striatum, consisting of the ventral and dorsal striatum. The dorsal striatum projects to the globus pallidus or to midbrain neurons in the SN, whereas the ventral striatum projects to the ventral pallidum or to the midbrain neurons of the VTA. Both the globus pallidus and the ventral pallidum also project to the midbrain neurons in the VTA−SN (for a review of basal ganglia circuitry, see DeLong and Wichmann, 2009). The basal ganglia circuitry is both heavily influenced by dopamine inputs and heavily influences dopamine neuron activity. Moreover, the anatomy of the circuit indicates a topographic gradient that moves from the more ventromedial aspects of the striatum to the more dorsolateral aspects, a gradient reflected in the topographical inputs and outputs between the striatum and the VTA−SN. Intriguingly, evidence suggests that information processing within the striatum occurs progressively along this ventromedial−dorsolateral gradient at least in part through a series of ascending loops (Haber et al., 2000; Ikemoto, 2007). Each part of the striatum innervates the midbrain cells that also target the part of the striatum creating a reciprocal loop. However, each part of the striatum also innervates the midbrain cells that target the striatum immediately dorsolateral to the part of the striatum innervating those cells. Thus, via these ascending reciprocal loops between the striatum and the midbrain dopamine cells, information processing can move from the NAshell to the most dorsolateral parts of the dorsal striatum.

Working Memory

The PFC is well known to be a critical mediator of working memory and, as noted, dopamine neurons in the VTA provide a significant innervation of the PFC. Evidence indicates that the PFC levels of dopamine dramatically influence PFC functioning and the resulting behavior during a working memory task (Brozoski et al., 1979; Sawaguchi and Goldman-Rakic, 1991). In general, it appears that dopamine levels in the PFC increase as the arousal levels of an organism shift from drowsy to alert to uncontrollably stressed. PFC functioning and an animal's success in a working memory task, however, follow an inverted-U function, in which moderate levels of dopamine increase an animal's working memory abilities but lower or higher levels impair working memory (Robbins and Arnsten, 2009), consistent with the idea that high levels of stress impair executive function and working memory abilities. Indeed, evidence indicates that microinjections of low doses of dopamine into the PFC enhance working memory performance, whereas higher doses impair such performance (Dent and Neill, 2012). However, in that study, the highest dose enhanced a stress-related behavior (escape from a water maze) suggesting that the function of dopamine in the PFC in modulating behavior depends on the circumstances and can facilitate appropriate behavior depending on the stress level of the animal. Orexin neurons in the hypothalamus appear to be critical for mediating the effects of moderate arousal (alert) on activity of dopamine neurons that project to the PFC (Saper et al., 2001; Vittoz et al., 2008). Stressful conditions, in contrast, lead to amygdalar activation that increases PFC dopamine levels (Goldstein et al., 1996).

Blockade of D1 receptors in the PFC impairs performance during a spatial working memory task in which monkeys must make oculomotor responses based on the position of a previously viewed cue (Sawaguchi and Goldman-Rakic, 1991). Recording studies from monkeys engaged in such a task demonstrate that activation of D1 receptors decreases the "noise" from inputs mediating the nonpreferred direction (Sawaguchi and Goldman-Rakic, 1994; Williams and Goldman-Rakic, 1995; Vijayraghavan et al., 2007). In general, it appears that optimal stimulation of D1 receptors in the PFC produces an enhanced "tuning" of the individual neurons, making them less responsive to nonpreferred inputs. However, under stress conditions, high levels of D1 receptor activation inhibit all inputs, both preferred and nonpreferred, leading to an overall diminution in working memory performance (Vijayraghavan et al., 2007). The beneficial effects of D1 receptor stimulation on working memory likely occur as a result of a reduction in the spontaneous firing of the relevant pyramidal cell and an increase in the excitability of the neuron, thus increasing the signal-to-noise ratio in these cells (Yang and Seamans, 1996). D2 receptors, in contrast, appear to play little role in the firing of the "delay cells" that are important for working memory tasks. However, activation of D2 receptors alters activity of response-related cells (Wang et al., 2004). It is likely that the D2 receptors in the PFC modulate the communication between the PFC and the motor-related structures in the actual motor performance of the task (Arnsten, 2011).

Although human studies are not as clear on the role for dopamine in the PFC during working memory due to the difficulties in conducting such work, the limited studies that exist suggest that PFC dopamine is playing a similar role in humans (Robbins and Arnsten, 2009). Indeed, recent imaging studies indicate that the dorsolateral PFC appears to be critical for the "context updating" involved in working memory tasks and that the SN and VTA show a phasic response to the context updating (D'Ardenne et al., 2012). Moreover, the activity in the SN and VTA, as measured using BOLD responses, positively correlated with activity in the dorsolateral PFC in a manner suggesting that the midbrain dopamine regions were responsible for modulating PFC activity during the working memory task.

Reinforcement Learning

Early hypotheses of the function of mesocorticolimbic dopamine focused on the idea that dopamine served as a positive hedonic signal (Wise et al., 1978), a belief still reflected in the lay media. However, evidence began to suggest that dopamine served as a type of reward signal, rather than a signal of "pleasure" (Schultz, 2000; Berridge, 2007; Salamone, 2007). Much of the evidence for this role for dopamine originated in studies using intracranial self-stimulation and drug self-administration. It became clear that the basic reinforcing properties of drugs of abuse depended on activation of dopamine receptors, especially those in the NA (McCutcheon et al., 2012). Indeed, blockade of such receptors reduces the reinforcing power of psychostimulants and nicotine (Woolverton and Virus, 1989; Corrigall et al., 1992; McGregor and Roberts, 1993).

Since the 1990s, there has been increasing evidence that dopamine's role in reinforcement learning is that of a reward-prediction error signal (Schultz, 1998, 2012; Stuber et al., 2008). Specifically, changes in dopamine neuron firing, and the resulting changes in dopamine release in target structures, signal violations of the organism's expectations regarding a reward. When a reward is given to an animal unexpectedly, dopamine neuron firing increases. However, when a conditioned stimulus, such as a tone, is presented prior to the presentation of the reward, a shift occurs over repeated trials, in which the increase in dopamine neuron firing occurs in response to the tone and eventually stops occurring in response to the reward itself. In this case, the reward is now being effectively signaled by the tone and, therefore, presentation of the reward is expected—i.e., it does not violate the animal's expectations. In contrast, because the animal does not know when the tone will come, presentation of the tone is a violation of expectations, leading to an increase in dopamine neuron firing. Further evidence of dopamine serving as a reward-prediction error signal comes from experiments in which the conditioned stimulus is presented but no reward follows. In this case, because the organism is expecting a reward but does not receive it, a decrease in dopamine neuron firing is observed during the period when the reward should occur.

Considerable evidence supports a critical role for dopamine, particularly the mesocorticolimbic pathway, in reinforcement/reward learning. Inactivation of the

VTA reduces the reinforcing power of medial forebrain bundle self-stimulation (Willick and Kokkinidis, 1995). Optogenetic stimulation of VTA dopamine neurons enhances the reinforcement of a food reward in an operant task (Adamantidis et al., 2011). Moreover, blockade of dopamine receptors in the NAcore prevents the acquisition of a Pavlovian approach task (Di Ciano et al., 2001). Similarly, blockade of both D1 and NMDA receptors in the NAcore prevents the acquisition of an instrumental learning task, pressing a lever for food (Smith-Roe and Kelley, 2000).

As noted in the anatomy of the basal ganglia, there is a ventromedial–dorsolateral gradient in the striatum in terms of dopamine inputs, as the VTA innervates the more ventromedial regions of the striatum (i.e., the NA) and the SNc innervates the more dorsolateral regions. However, this gradient is also believed to reflect a gradient in functioning in terms of learning and memory. It appears that initial reinforcement learning is mediated by the NAshell and its dopamine inputs, but as the contingencies become better learned, the animal's behavior shifts to goal-directed behavior that is mediated by the NAcore. As this behavior becomes even better learned and eventually overlearned, the dorsolateral striatum and its dopamine inputs begin to mediate the behavior (Belin et al., 2009). Indeed, evidence suggests that habit learning, or stimulus-response learning, requires activity in the dorsal striatum (Packard and McGaugh, 1996; Packard and Knowlton, 2002). For example, posttraining intra-dorsal striatum blockade of dopamine receptors impairs retention of a habit-based radial-arm maze task (Legault et al., 2006). Similarly, lesions of the nigrostriatal pathway prevent the shift in an instrumental task from goal-directed behavior to habit-based, stimulus-response behavior (Faure et al., 2005). Thus, the evidence indicates that the dopamine inputs to the striatum are crucial for the development of the relevant motivated behaviors, from reinforcement learning to habit learning, in a manner that corresponds to the topographically determined functions of the striatum.

Dopamine and Neural Plasticity

Studies indicate that dopamine and the subsequent activation of dopamine receptors play crucial roles in the development of synaptic plasticity, particularly as assessed in long-term potentiation (LTP) and long-term depression (LTD). Within the CA1 region of the hippocampus, evidence indicates that administration of a D1 agonist induces LTP in a cAMP-dependent manner, whereas administration of a D2 agonist has no effect (Huang and Kandel, 1995), though evidence suggests that this dopamine effect on LTP also depends on NMDA receptors and the initial test pulses (Navakkode et al., 2007; Lisman et al., 2011). Moreover, it appears that dopamine is required for late LTP, but not early LTP (Frey et al., 1990), an effect suggesting the importance of dopamine in promoting the protein synthesis necessary for late LTP (Lisman et al., 2011). Within the BLA, D1 receptor activation induces LTP with a subthreshold induction protocol, whereas D1 receptor blockade prevents the induction of LTP (Li et al., 2011). In the PFC, spike-timing induced LTP requires activation of both D1 and D2 receptors (Xu and Yao, 2010). It appears that activation of postsynaptic D1 receptors permits the induction of LTP within longer-than-normal timing intervals, whereas the D2 receptors act to inhibit local GABA neurons, thus disinhibiting the pyramidal cells (Xu and Yao, 2010). Thus, evidence from studies on these glutamatergic structures that are major targets of the mesocorticolimbic inputs demonstrates the important role of dopamine in modulating changes in synaptic efficacy that likely underlie memory processes.

Similarly, studies on the striatum indicate that dopamine influences the development of plasticity at glutamatergic (likely cortical) inputs to the striatal cells. Blockade of D1 receptors prevents induction of LTP at corticostriatal synapses (Calabresi et al., 2000; Kerr and Wickens, 2001), whereas D2 receptor blockade appears to block LTD at these synapses (Tang et al., 2001; Kreitzer and Malenka, 2005). The activation of D2 receptors appears to lead to increased production of endocannabinoids that activate presynaptic receptors, leading to the reduction in synaptic efficacy (Kreitzer and Malenka, 2008). Considering that striatal neurons express either D1 or D2 receptors, whether dopamine is necessary in inducing LTP or LTD at synapses onto the different striatal neurons is not clear, but evidence suggests that the necessity of dopamine depends on the receptor subtype expressed by the cell (Kreitzer and Malenka, 2008). It appears that a key mechanism of dopamine's effects on synaptic plasticity comes through the regulation of glutamatergic α-Amino-3-hydroxy-5-methyl-4-isoxazolepropionic acid (AMPA) receptors. In NA neurons, D1 receptor activation increases phosphorylation of the GluA1 receptor subunit at the PKA phosphorylation site, which produces an enhanced AMPA receptor current, and increases the surface expression of the GluA1 subunit (Wolf et al., 2003). As a result, dopamine can directly facilitate glutamatergic synaptic transmission in the NA, thus enabling the alteration of circuits mediating motivated behavior. Indeed, LTP induced by stimulation of glutamatergic inputs to the NA is blocked by D1, but not D2, receptor blockade (Schotanus and Chergui, 2008). However, the precise roles of the receptor subtypes in mediating different forms of plasticity in the striatal subcompartments are still being determined and, importantly, the functional relevance of these findings remains to be elucidated. Nonetheless, from the work presented here, it is clear that dopamine influences long-term plasticity mechanisms and is even required for some of them.

Dopamine and Memory Consolidation

Although, as noted, there is substantial evidence for dopamine neuron activity serving as a reward-prediction error signal, studies indicate that dopamine neuron activity, and subsequent release of dopamine in target structures, also signals nonreward events (McCutcheon et al., 2012). In particular, novel stimuli induce burst firing in dopamine cells and subsequent release of dopamine in target regions (Ljungberg et al., 1992; Rebec, 1998; De Leonibus et al., 2006). Moreover, and of importance in this section, studies have demonstrated that dopamine neurons also fire in response to aversive stimuli as well as cues that signal aversive stimuli (Matsumoto and Hikosaka, 2009; Bromberg-Martin et al., 2010). Indeed, the evidence suggests that different subpopulations of dopamine neurons respond to rewarding versus aversive stimuli and that at least some of the neurons increase firing to both kinds of stimuli (Wang and Tsien, 2011). As a result of these findings, as well as the behavioral ones described later, it is likely that, on a more general level, dopamine neuron activity and release of dopamine in target structures is involved in signaling motivationally salient stimuli, whether those stimuli are simply novel, rewarding, or aversive. Dopamine can then serve, in the target structure, to modulate plasticity regarding the animal's behavior toward these stimuli, thus promoting memories for the event.

Indeed, on a behavioral level, considerable evidence indicates that the dopaminergic system is involved in the acquisition/consolidation for a variety of nonreward-related tasks. In Pavlovian footshock-based tasks, dopamine activity in the BLA, the hippocampus, and the NA appear to be critical for normal memory consolidation. Dopamine-deficient mice are impaired in their long-term memory for fear-potentiated startle, an effect that is reversed by selective restoration of dopamine to both the BLA and the NA (Fadok et al., 2009, 2010). Pretraining intra-amygdala administration of a D1 or D2 receptor antagonist prior to a Pavlovian fear conditioning session impairs retention (Greba and Kokkinidis, 2000; Guarraci et al., 2000; Greba et al., 2001), whereas intra-amygdala administration of a D1 receptor agonist enhances retention in the same task (Guarraci et al., 1999). Posttraining intra-BLA or intra-NAshell administration of dopamine enhances the retention for inhibitory avoidance, whereas posttraining intra-BLA administration of a D1 or D2 receptor antagonist impairs retention (Lalumiere et al., 2004, 2005). The memory enhancement with dopamine infusions into either structure, however, requires concurrent activation of dopamine receptors in the other structure (Lalumiere et al., 2005), suggesting the critical nature of dopamine inputs to both the BLA and the NA for normal memory consolidation. Electrophysiology studies indicate that dopamine plays a crucial role in enhancing the neuronal excitability of BLA neurons that follows an odor-footshock pairing (Rosenkranz and Grace, 2002), confirming the idea that the increased BLA activity following training that modulates memory consolidation is driven, at least in part, by the actions of dopamine.

As noted in the anatomy section, the hippocampus receives a significant dopaminergic input, which appears to be critically involved in memory processes as well. Knockdown of D1 receptors in the hippocampus of mice impairs learning in a variety of associative learning tasks and impairs the ability to induce LTP in the hippocampus, indicating the importance of D1 receptors in neuroplasticity mechanisms (Ortiz et al., 2010). Blockade of D1 receptors in the dorsal hippocampus 12 h after training has no effect on retention of inhibitory avoidance 2 days later but disrupts retention 7 and 14 days later (Rossato et al., 2009). Intriguingly, D1 receptor blockade in the dorsal hippocampus immediately after training has no effect on retention, suggesting that the activation of D1 receptors in the hippocampus is critical for the persistence of the long-term memory. Moreover, it suggests that dopaminergic signaling well after the learning event is playing a crucial role regulating hippocampal activity and plasticity. Indeed, blockade of VTA NMDA receptors also prevents the persistence of the long-term memory, an effect that is reversed by administration of a D1 receptor agonist into the hippocampus (Rossato et al., 2009).

Taste memory appears to involve dopamine in the insular cortex. Following a conditioned taste aversion paradigm, there is a significant increase in dopamine levels in the insular cortex approximately 40 min after the lithium chloride injection, an effect blocked by BLA inactivation, suggesting the importance of the BLA in regulating the dopamine system during the consolidation period (Guzman-Ramos et al., 2010). Intriguingly, blockade of D1 receptors in the insular cortex alone had no effect on retention of the conditioned taste aversion but potentiated the impairing effects of concurrent administration of an NMDA receptor antagonist into the insular cortex (Guzman-Ramos et al., 2010).

Studies examining the extinction of learned behavior demonstrate a role for dopamine in such learning as well. Activation of dopamine receptors within a variety of structures, including the ventromedial PFC, the BLA, and the dorsal hippocampus, appears to be necessary for normal extinction for fear conditioning and inhibitory avoidance (Mueller et al., 2010; Fiorenza et al., 2012). Systemic administration of a D1 receptor antagonist given immediately after retrieval sessions impairs the consolidation of extinction of cocaine conditioned place preference (Fricks-Gleason et al., 2012). Blockade of D4 receptors in the medial PFC impair retention of extinction learning in a fear conditioning task (Pfeiffer and Fendt, 2006). Loss of dopamine in the PFC impairs extinction learning for fear conditioning (Fernandez Espejo, 2003). Based on these findings, it appears that dopamine plays a key role in the consolidation of memories across a variety of different types of learning and in a variety of brain structures, suggesting that dopamine influences neuroplasticity within the structure. Moreover, these findings suggest that dopamine's role in learning and memory is determined by the function of the innervated structure.

The results from the animal behavior studies and the synaptic plasticity experiments suggest that dopamine is particularly critical for the conversion of memories into long-term storage. Evidence from human imaging studies appears to confirm this. Administration of L-DOPA, a precursor molecule for the synthesis of dopamine, enhanced retention for images viewed by elderly patients, regardless of how strongly the images were encoded by the hippocampus (Chowdhury et al., 2012). The enhancement followed an inverted-U dose−response relationship. However, the enhancement only occurred for retention tests given after 6 h following initial encoding and not for tests given 2 h after encoding, further supporting the idea that dopamine is critical for long-term memory consolidation processes.

Conclusions

The findings presented in this chapter indicate that the mesocorticolimbic and nigrostriatal dopamine systems are critically involved in learning and memory processes. The midbrain dopamine cells provide significant innervations of the striatal circuitry, limbic systems, and cortical regions that interact with those systems, thus positioning the dopamine system to be a crucial modulator of plasticity and memory related to motivationally relevant stimuli. Indeed, evidence suggests a key role for dopamine in learning related not only to reward but also to novelty and aversive stimuli. In particular, the dopamine innervation of the striatum appears to influence memory processes related to reinforcement learning, goal-directed behavior, and habit, whereas its innervation of limbic regions, such as the hippocampus and amygdala, appear to be more involved in modulating memory consolidation for motivationally significant stimuli in general. Moreover, evidence suggests that the dopamine projections to the PFC have a particular optimizing prefrontal functioning in order to undergird working memory processes. Thus, dopamine inputs throughout the forebrain regulate a variety of memory processes, depending on the task under study as well as the overall function of the innervated structure.

References

Adamantidis, A.R., Tsai, H.C., Boutrel, B., Zhang, F., Stuber, G.D., Budygin, E.A., et al., 2011. Optogenetic interrogation of dopaminergic modulation of the multiple phases of reward-seeking behavior. J. Neurosci. 31, 10829—10835.

Anden, N.E., Carlsson, A., Dahlstroem, A., Fuxe, K., Hillarp, N.A., Larsson, K., 1964. Demonstration and mapping out of nigro-neostriatal dopamine neurons. Life Sci. 3, 523—530.

Andersen, P.H., Gingrich, J.A., Bates, M.D., Dearry, A., Falardeau, P., Senogles, S.E., et al., 1990. Dopamine receptor subtypes: beyond the D1/D2 classification. Trends Pharmacol. Sci. 11, 231—236.

Arnsten, A.F., 2011. Catecholamine influences on dorsolateral prefrontal cortical networks. Biol. Psychiatry. 69, e89—e99.

Asan, E., 1998. The catecholaminergic innervation of the rat amygdala. Adv. Anat. Embryol. Cell Biol. 142, 1—118.

Beaulieu, J.M., Gainetdinov, R.R., 2011. The physiology, signaling, and pharmacology of dopamine receptors. Pharmacol. Rev. 63, 182—217.

Belin, D., Jonkman, S., Dickinson, A., Robbins, T.W., Everitt, B.J., 2009. Parallel and interactive learning processes within the basal ganglia: relevance for the understanding of addiction. Behav. Brain Res. 199, 89—102.

Berridge, K.C., 2007. The debate over dopamine's role in reward: the case for incentive salience. Psychopharmacology. 191, 391—431.

Bolam, J.P., Smith, Y., 1990. The GABA and substance P input to dopaminergic neurones in the substantia nigra of the rat. Brain Res. 529, 57—78.

Bromberg-Martin, E.S., Matsumoto, M., Hikosaka, O., 2010. Dopamine in motivational control: rewarding, aversive, and alerting. Neuron. 68, 815—834.

Brozoski, T.J., Brown, R.M., Rosvold, H.E., Goldman, P.S., 1979. Cognitive deficit caused by regional depletion of dopamine in prefrontal cortex of rhesus monkey. Science. 205, 929—932.

Calabresi, P., Gubellini, P., Centonze, D., Picconi, B., Bernardi, G., Chergui, K., et al., 2000. Dopamine and cAMP-regulated phosphoprotein 32 kDa controls both striatal long-term depression and long-term potentiation, opposing forms of synaptic plasticity. J. Neurosci. 20, 8443—8451.

Carlsson, A., 1987. Perspectives on the discovery of central monoaminergic neurotransmission. Annu. Rev. Neurosci. 10, 19–40.

Carlsson, A., Lindqvist, M., Magnusson, T., 1957. 3,4-Dihydroxyphenylalanine and 5-hydroxytryptophan as reserpine antagonists. Nature. 180, 1200.

Carlsson, A., Falck, B., Hillarp, N.A., 1962. Cellular localization of brain monoamines. Acta Physiol. Scand. Suppl. 56, 1–28.

Chowdhury, R., Guitart-Masip, M., Bunzeck, N., Dolan, R.J., Duzel, E., 2012. Dopamine modulates episodic memory persistence in old age. J. Neurosci. 32, 14193–14204.

Ciliax, B.J., Heilman, C., Demchyshyn, L.L., Pristupa, Z.B., Ince, E., Hersch, S.M., et al., 1995. The dopamine transporter: immunochemical characterization and localization in brain. J. Neurosci. 15, 1714–1723.

Corrigall, W.A., Franklin, K.B., Coen, K.M., Clarke, P.B., 1992. The mesolimbic dopaminergic system is implicated in the reinforcing effects of nicotine. Psychopharmacology. 107, 285–289.

D'Ardenne, K., Eshel, N., Luka, J., Lenartowicz, A., Nystrom, L.E., Cohen, J.D., 2012. Role of prefrontal cortex and the midbrain dopamine system in working memory updating. Proc. Natl. Acad. Sci. USA. 109, 19900–19909.

Dahlstroem, A., Fuxe, K., 1964. Evidence for the existence of monoamine-containing neurons in the central nervous system. I. Demonstration of monoamines in the cell bodies of brain stem neurons. Acta Physiol. Scand. Suppl. Suppl. 232, 231–255.

De Leonibus, E., Verheij, M.M., Mele, A., Cools, A., 2006. Distinct kinds of novelty processing differentially increase extracellular dopamine in different brain regions. Eur. J. Neurosci. 23, 1332–1340.

DeLong, M., Wichmann, T., 2009. Update on models of basal ganglia function and dysfunction. Parkinsonism Relat. Disord. 15 (Suppl. 3), S237–S240.

Dent, M.F., Neill, D.B., 2012. Dose-dependent effects of prefrontal dopamine on behavioral state in rats. Behav. Neurosci. 126, 620–639.

Descarries, L., Berube-Carriere, N., Riad, M., Bo, G.D., Mendez, J.A., Trudeau, L.E., 2008. Glutamate in dopamine neurons: synaptic versus diffuse transmission. Brain Res. Rev. 58, 290–302.

Di Ciano, P., Cardinal, R.N., Cowell, R.A., Little, S.J., Everitt, B.J., 2001. Differential involvement of NMDA, AMPA/kainate, and dopamine receptors in the nucleus accumbens core in the acquisition and performance of Pavlovian approach behavior. J. Neurosci. 21, 9471–9477.

Fadok, J.P., Dickerson, T.M., Palmiter, R.D., 2009. Dopamine is necessary for cue-dependent fear conditioning. J. Neurosci. 29, 11089–11097.

Fadok, J.P., Darvas, M., Dickerson, T.M., Palmiter, R.D., 2010. Long-term memory for Pavlovian fear conditioning requires dopamine in the nucleus accumbens and basolateral amygdala. PLoS One. 5, e12751.

Fallon, J.H., Loughlin, S.E., 1995. Substantia nigra. In: Paxinos, G. (Ed.), The Rat Nervous System, second ed. Academic Press, San Diego, CA.

Fallon, J.H., Koziell, D.A., Moore, R.Y., 1978. Catecholamine innervation of the basal forebrain. II. Amygdala, suprarhinal cortex and entorhinal cortex. J. Comp. Neurol. 180, 509–532.

Faure, A., Haberland, U., Conde, F., El Massioui, N., 2005. Lesion to the nigrostriatal dopamine system disrupts stimulus-response habit formation. J. Neurosci. 25, 2771–2780.

Felder, C.C., Jose, P.A., Axelrod, J., 1989. The dopamine-1 agonist, SKF 82526, stimulates phospholipase-C activity independent of adenylate cyclase. J. Pharmacol. Exp. Ther. 248, 171–175.

Fernandez Espejo, E., 2003. Prefrontocortical dopamine loss in rats delays long-term extinction of contextual conditioned fear, and reduces social interaction without affecting short-term social interaction memory. Neuropsychopharmacology. 28, 490−498.

Fiorenza, N.G., Rosa, J., Izquierdo, I., Myskiw, J.C., 2012. Modulation of the extinction of two different fear-motivated tasks in three distinct brain areas. Behav. Brain Res. 232, 210−216.

Fonnum, F., Gottesfeld, Z., Grofova, I., 1978. Distribution of glutamate decarboxylase, choline acetyl-transferase and aromatic amino acid decarboxylase in the basal ganglia of normal and operated rats. Evidence for striatopallidal, striatoentopeduncular and striatonigral GABAergic fibres. Brain Res. 143, 125−138.

Frey, U., Schroeder, H., Matthies, H., 1990. Dopaminergic antagonists prevent long-term maintenance of posttetanic LTP in the CA1 region of rat hippocampal slices. Brain Res. 522, 69−75.

Fricks-Gleason, A.N., Khalaj, A.J., Marshall, J.F., 2012. Dopamine D1 receptor antagonism impairs extinction of cocaine-cue memories. Behav. Brain Res. 226, 357−360.

Gasbarri, A., Sulli, A., Packard, M.G., 1997. The dopaminergic mesencephalic projections to the hippocampal formation in the rat. Prog. Neuropsychopharmacol. Biol. Psychiatry. 21, 1−22.

Gerfen, C.R., Herkenham, M., Thibault, J., 1987. The neostriatal mosaic: II. Patch- and matrix-directed mesostriatal dopaminergic and non-dopaminergic systems. J. Neurosci. 7, 3915−3934.

Goldstein, L.E., Rasmusson, A.M., Bunney, B.S., Roth, R.H., 1996. Role of the amygdala in the coordination of behavioral, neuroendocrine, and prefrontal cortical monoamine responses to psychological stress in the rat. J. Neurosci. 16, 4787−4798.

Greba, Q., Kokkinidis, L., 2000. Peripheral and intraamygdalar administration of the dopamine D1 receptor antagonist SCH 23390 blocks fear-potentiated startle but not shock reactivity or the shock sensitization of acoustic startle. Behav. Neurosci. 114, 262−272.

Greba, Q., Gifkins, A., Kokkinidis, L., 2001. Inhibition of amygdaloid dopamine D2 receptors impairs emotional learning measured with fear-potentiated startle. Brain Res. 899, 218−226.

Grillner, P., Mercuri, N.B., 2002. Intrinsic membrane properties and synaptic inputs regulating the firing activity of the dopamine neurons. Behav. Brain Res. 130, 149−169.

Guarraci, F.A., Frohardt, R.J., Kapp, B.S., 1999. Amygdaloid D1 dopamine receptor involvement in Pavlovian fear conditioning. Brain Res. 827, 28−40.

Guarraci, F.A., Frohardt, R.J., Falls, W.A., Kapp, B.S., 2000. The effects of intra-amygdaloid infusions of a D2 dopamine receptor antagonist on Pavlovian fear conditioning. Behav. Neurosci. 114, 647−651.

Guzman-Ramos, K., Osorio-Gomez, D., Moreno-Castilla, P., Bermudez-Rattoni, F., 2010. Off-line concomitant release of dopamine and glutamate involvement in taste memory consolidation. J. Neurochem. 114, 226−236.

Haber, S.N., Fudge, J.L., McFarland, N.R., 2000. Striatonigrostriatal pathways in primates form an ascending spiral from the shell to the dorsolateral striatum. J. Neurosci. 20, 2369−2382.

Hernandez-Lopez, S., Tkatch, T., Perez-Garci, E., Galarraga, E., Bargas, J., Hamm, H., et al., 2000. D2 dopamine receptors in striatal medium spiny neurons reduce L-type Ca2 + currents and excitability via a novel PLC[beta]1-IP3-calcineurin-signaling cascade. J. Neurosci. 20, 8987−8995.

Hnasko, T.S., Chuhma, N., Zhang, H., Goh, G.Y., Sulzer, D., Palmiter, R.D., et al., 2010. Vesicular glutamate transport promotes dopamine storage and glutamate corelease in vivo. Neuron. 65, 643−656.

Huang, Y.Y., Kandel, E.R., 1995. D1/D5 receptor agonists induce a protein synthesis-dependent late potentiation in the CA1 region of the hippocampus. Proc. Natl. Acad. Sci. USA. 92, 2446–2450.

Ikemoto, S., 2007. Dopamine reward circuitry: two projection systems from the ventral midbrain to the nucleus accumbens–olfactory tubercle complex. Brain Res. Rev. 56, 27–78.

Kerr, J.N., Wickens, J.R., 2001. Dopamine D-1/D-5 receptor activation is required for long-term potentiation in the rat neostriatum in vitro. J. Neurophysiol. 85, 117–124.

Kreitzer, A.C., Malenka, R.C., 2005. Dopamine modulation of state-dependent endocannabinoid release and long-term depression in the striatum. J. Neurosci. 25, 10537–10545.

Kreitzer, A.C., Malenka, R.C., 2008. Striatal plasticity and basal ganglia circuit function. Neuron. 60, 543–554.

Lalumiere, R.T., Nguyen, L.T., McGaugh, J.L., 2004. Post-training intrabasolateral amygdala infusions of dopamine modulate consolidation of inhibitory avoidance memory: involvement of noradrenergic and cholinergic systems. Eur. J. Neurosci. 20, 2804–2810.

Lalumiere, R.T., Nawar, E.M., McGaugh, J.L., 2005. Modulation of memory consolidation by the basolateral amygdala or nucleus accumbens shell requires concurrent dopamine receptor activation in both brain regions. Learn. Mem. 12, 296–301.

Legault, G., Smith, C.T., Beninger, R.J., 2006. Post-training intra-striatal scopolamine or flupenthixol impairs radial maze learning in rats. Behav. Brain Res. 170, 148–155.

Li, C., Dabrowska, J., Hazra, R., Rainnie, D.G., 2011. Synergistic activation of dopamine D1 and TrkB receptors mediate gain control of synaptic plasticity in the basolateral amygdala. PLoS One. 6, e26065.

Li, X., Qi, J., Yamaguchi, T., Wang, H.L., Morales, M., 2013. Heterogeneous composition of dopamine neurons of the rat A10 region: molecular evidence for diverse signaling properties. Brain Struct. Funct. 218, pp. 1159–1176.

Lisman, J., Grace, A.A., Duzel, E., 2011. A neoHebbian framework for episodic memory; role of dopamine-dependent late LTP. Trends Neurosci. 34, 536–547.

Ljungberg, T., Apicella, P., Schultz, W., 1992. Responses of monkey dopamine neurons during learning of behavioral reactions. J. Neurophysiol. 67, 145–163.

Matsumoto, M., Hikosaka, O., 2009. Two types of dopamine neuron distinctly convey positive and negative motivational signals. Nature. 459, 837–841.

McCutcheon, J.E., Ebner, S.R., Loriaux, A.L., Roitman, M.F., 2012. Encoding of aversion by dopamine and the nucleus accumbens. Front Neurosci. 6, 137.

McGregor, A., Roberts, D.C., 1993. Dopaminergic antagonism within the nucleus accumbens or the amygdala produces differential effects on intravenous cocaine self-administration under fixed and progressive ratio schedules of reinforcement. Brain Res. 624, 245–252.

Missale, C., Nash, S.R., Robinson, S.W., Jaber, M., Caron, M.G., 1998. Dopamine receptors: from structure to function. Physiol. Rev. 78, 189–225.

Mueller, D., Bravo-Rivera, C., Quirk, G.J., 2010. Infralimbic D2 receptors are necessary for fear extinction and extinction-related tone responses. Biol. Psychiatry. 68, 1055–1060.

Nair-Roberts, R.G., Chatelain-Badie, S.D., Benson, E., White-Cooper, H., Bolam, J.P., Ungless, M.A., 2008. Stereological estimates of dopaminergic, GABAergic and glutamatergic neurons in the ventral tegmental area, substantia nigra and retrorubral field in the rat. Neuroscience. 152, 1024–1031.

Navakkode, S., Sajikumar, S., Frey, J.U., 2007. Synergistic requirements for the induction of dopaminergic D1/D5-receptor-mediated LTP in hippocampal slices of rat CA1 in vitro. Neuropharmacology. 52, 1547–1554.

Niznik, H.B., Van Tol, H.H., 1992. Dopamine receptor genes: new tools for molecular psychiatry. J. Psychiatry Neurosci. 17, 158–180.

Ortiz, O., Delgado-Garcia, J.M., Espadas, I., Bahi, A., Trullas, R., Dreyer, J.L., et al., 2010. Associative learning and CA3-CA1 synaptic plasticity are impaired in D1R null, Drd1a-/- mice and in hippocampal siRNA silenced Drd1a mice. J. Neurosci. 30, 12288–12300.

Packard, M.G., Knowlton, B.J., 2002. Learning and memory functions of the basal ganglia. Annu. Rev. Neurosci. 25, 563–593.

Packard, M.G., McGaugh, J.L., 1996. Inactivation of hippocampus or caudate nucleus with lidocaine differentially affects expression of place and response learning. Neurobiol. Learn. Mem. 65, 65–72.

Pfeiffer, U.J., Fendt, M., 2006. Prefrontal dopamine D4 receptors are involved in encoding fear extinction. Neuroreport. 17, 847–850.

Rebec, G.V., 1998. Real-time assessments of dopamine function during behavior: single-unit recording, iontophoresis, and fast-scan cyclic voltammetry in awake, unrestrained rats. Alcohol. Clin. Exp. Res. 22, 32–40.

Robbins, T.W., Arnsten, A.F., 2009. The neuropsychopharmacology of fronto-executive function: monoaminergic modulation. Annu. Rev. Neurosci. 32, 267–287.

Rosenkranz, J.A., Grace, A.A., 2002. Dopamine-mediated modulation of odour-evoked amygdala potentials during Pavlovian conditioning. Nature. 417, 282–287.

Rossato, J.I., Bevilaqua, L.R., Izquierdo, I., Medina, J.H., Cammarota, M., 2009. Dopamine controls persistence of long-term memory storage. Science. 325, 1017–1020.

Salamone, J.D., 2007. Functions of mesolimbic dopamine: changing concepts and shifting paradigms. Psychopharmacology. 191, 389.

Saper, C.B., Chou, T.C., Scammell, T.E., 2001. The sleep switch: hypothalamic control of sleep and wakefulness. Trends Neurosci. 24, 726–731.

Sawaguchi, T., Goldman-Rakic, P.S., 1991. D1 dopamine receptors in prefrontal cortex: involvement in working memory. Science. 251, 947–950.

Sawaguchi, T., Goldman-Rakic, P.S., 1994. The role of D1-dopamine receptor in working memory: local injections of dopamine antagonists into the prefrontal cortex of rhesus monkeys performing an oculomotor delayed-response task. J. Neurophysiol. 71, 515–528.

Schotanus, S.M., Chergui, K., 2008. Dopamine D1 receptors and group I metabotropic glutamate receptors contribute to the induction of long-term potentiation in the nucleus accumbens. Neuropharmacology. 54, 837–844.

Schultz, W., 1998. Predictive reward signal of dopamine neurons. J. Neurophysiol. 80, 1–27.

Schultz, W., 2000. Multiple reward signals in the brain. Nat. Rev. Neurosci. 1, 199–207.

Schultz, W., 2013. Updating dopamine reward signals. Curr. Opin. Neurobiol. 23, pp. 229–238.

Smith, Y., Bolam, J.P., 1990. The output neurones and the dopaminergic neurones of the substantia nigra receive a GABA-containing input from the globus pallidus in the rat. J. Comp. Neurol. 296, 47–64.

Smith-Roe, S.L., Kelley, A.E., 2000. Coincident activation of NMDA and dopamine D1 receptors within the nucleus accumbens core is required for appetitive instrumental learning. J. Neurosci. 20, 7737–7742.

Stuber, G.D., Klanker, M., de Ridder, B., Bowers, M.S., Joosten, R.N., Feenstra, M.G., et al., 2008. Reward-predictive cues enhance excitatory synaptic strength onto midbrain dopamine neurons. Science. 321, 1690–1692.

Takada, M., 1990. The A11 catecholamine cell group: another origin of the dopaminergic innervation of the amygdala. Neurosci. Lett. 118, 132–135.

Tang, K., Low, M.J., Grandy, D.K., Lovinger, D.M., 2001. Dopamine-dependent synaptic plasticity in striatum during *in vivo* development. Proc. Natl. Acad. Sci. USA. 98, 1255–1260.

Ungerstedt, U., 1971. Stereotaxic mapping of the monoamine pathways in the rat brain. Acta Physiol. Scand. Suppl. 367, 1–48.

Vijayraghavan, S., Wang, M., Birnbaum, S.G., Williams, G.V., Arnsten, A.F., 2007. Inverted-U dopamine D1 receptor actions on prefrontal neurons engaged in working memory. Nat. Neurosci. 10, 376–384.

Vittoz, N.M., Schmeichel, B., Berridge, C.W., 2008. Hypocretin/orexin preferentially activates caudomedial ventral tegmental area dopamine neurons. Eur. J. Neurosci. 28, 1629–1640.

Wang, D.V., Tsien, J.Z., 2011. Convergent processing of both positive and negative motivational signals by the VTA dopamine neuronal populations. PLoS One. 6, e17047.

Wang, M., Vijayraghavan, S., Goldman-Rakic, P.S., 2004. Selective D2 receptor actions on the functional circuitry of working memory. Science. 303, 853–856.

Wayment, H.K., Schenk, J.O., Sorg, B.A., 2001. Characterization of extracellular dopamine clearance in the medial prefrontal cortex: role of monoamine uptake and monoamine oxidase inhibition. J. Neurosci. 21, 35–44.

Williams, G.V., Goldman-Rakic, P.S., 1995. Modulation of memory fields by dopamine D1 receptors in prefrontal cortex. Nature. 376, 572–575.

Willick, M.L., Kokkinidis, L., 1995. The effects of ventral tegmental administration of GABAA, GABAB and NMDA receptor agonists on medial forebrain bundle self-stimulation. Behav. Brain Res. 70, 31–36.

Wise, R.A., Spindler, J., deWit, H., Gerberg, G.J., 1978. Neuroleptic-induced "anhedonia" in rats: pimozide blocks reward quality of food. Science. 201, 262–264.

Wolf, M.E., Mangiavacchi, S., Sun, X., 2003. Mechanisms by which dopamine receptors may influence synaptic plasticity. Ann. NY Acad. Sci. 1003, 241–249.

Woolverton, W.L., Virus, R.M., 1989. The effects of a D1 and a D2 dopamine antagonist on behavior maintained by cocaine or food. Pharmacol. Biochem. Behav. 32, 691–697.

Xu, T.X., Yao, W.D., 2010. D1 and D2 dopamine receptors in separate circuits cooperate to drive associative long-term potentiation in the prefrontal cortex. Proc. Natl. Acad. Sci. USA. 107, 16366–16371.

Yamaguchi, T., Wang, H.L., Li, X., Ng, T.H., Morales, M., 2011. Mesocorticolimbic glutamatergic pathway. J. Neurosci. 31, 8476–8490.

Yang, C.R., Seamans, J.K., 1996. Dopamine D1 receptor actions in layers V-VI rat prefrontal cortex neurons *in vitro*: modulation of dendritic-somatic signal integration. J. Neurosci. 16, 1922–1935.

6 Unpacking Memory Processes: Using the Attribute Model to Design Optimal Memory Tests for Rodent Models

Michael R. Hunsaker

Department of Psychiatry and Behavioral Sciences and MIND Institute,
University of California, Davis Medical Center, Sacramento, CA

Introduction

An important consideration in the study of mouse models of any genetic disease is how well the behavior of the mouse serves to actually model the behaviors present in the clinical population being modeled. In order to optimally address this concern is to explicitly develop a behavioral endophenotype of the mouse model in which the mouse is explicitly tested using tasks rigorously designed to explicitly test the cognitive domains affected in the clinical population. To achieve this goal, it is important not only to apply standard behavioral tasks to mouse models of genetic disorders, but also to directly evaluate brain function across all cognitive domains. Furthermore, it is critical for behavioral genetics laboratories to interact with clinical research laboratories to develop comprehensive behavioral endophenotypes for the disorder being modeled. The strength of behavioral genetics and the mouse models is the ability to apply behavioral paradigms known to be subserved by known anatomical loci not only to determine the behavioral phenotype of the model, but also to elucidate candidate brain regions affected by the mutation (Robbins et al., 2012). When research into mouse models of genetic disorders emphasizes patterns of mnemonic strengths and weaknesses across domains (i.e., the behavioral endophenotype), the results will not only directly model the disorder being studied, but also serve both as risk prodrome for disease onset or progression and as outcome measures that can be applied by the clinic in interventional studies.

In this chapter, I will describe a process wherein the behavioral geneticist can identify and select behavioral tests for a mutant mouse with minimal difficulty. In doing so I will (1) describe a comprehensive model of memory processing based on multiple memory systems and parallel processing, (2) discuss the utility of this considering this model prior to designing and selecting behavioral experiments for

Identification of Neural Markers Accompanying Memory. DOI: http://dx.doi.org/10.1016/B978-0-12-408139-0.00006-7

a genetic model, and (3) provide an example from my own work with the CGG KI mouse model of the fragile X premutation to illustrate an example of using the attribute model to guide the development of a behavioral task battery for the mouse model.

Attribute Model

Memory Systems

At the level of processing, the event-based memory system provides for temporary representations of incoming data concerning the present (i.e., online processing), with an emphasis upon data and events that are usually personal and that occur within specific external and internal contexts. The emphasis is upon the processing of new and current information. During initial learning, emphasis is placed on the event-based memory system, which will continue to be of importance even after initial learning in situations where unique or novel trial information needs to be remembered (Hunsaker, 2012a,b; Hunsaker and Kesner, 2008, 2013; Kesner and Hunsaker, 2010; Kesner, 2013). This system is akin to episodic memory (Tulving, 1983) and some aspects of declarative memory (Squire, 1994) as formulated to describe research using human subjects.

The knowledge-based memory system provides for more permanent representations of previously stored information in long-term memory and can be thought of as one's general knowledge of the world. The knowledge-based memory system would tend to be of greater importance after a task has been learned or given that the situation has become invariant and/or familiar. This system is akin to semantic memory and can be summarized as retrieval and consolidation processes (Tulving, 1983).

The rule-based memory system receives information from the event- and knowledge-based systems and integrates the information by applying rules and strategies for subsequent action (Churchwell and Kesner, 2011; Churchwell et al., 2009, 2010; Cohen and O'Reilly, 1996; Kesner and Churchwell, 2011). In most situations, however, one would expect a contribution of all three systems with a varying proportion of involvement of one relative to the other depending primarily upon the demands of the task being performed.

Specific Attributes

The three memory systems are composed of the same forms, domains, or attributes of memory processes. Even though there could be many attributes, the most important attributes in the attribute model as presently formulated are space, time, response, sensory-perception, and reward value (affect). In humans, a language attribute is also added.

A spatial (space) attribute within this framework involves memory representations of places or relationships between places. It is exemplified by the ability to encode and remember spatial maps and to localize stimuli in external space.

Table 6.1 Description of the Processes Performed by Different Memory Systems Used in the Attribute Theory as Applicable to Research Using Rodents

	Event Based	Knowledge Based	Rule Based
Encoding	• Pattern separation • Transient representations • Short-term memory • Intermediate-term memory	• Selective attention • Associated with permanent memory representations • Perceptual memory	• Strategy selection • Rule maintenance
Retrieval	• Consolidation • Pattern completion	• Long-term memory • Retrieval based on flexibility and action	• Short-term working memory

Memory representations of the spatial attribute can be further subdivided into specific spatial features including allocentric spatial distance, egocentric spatial distance, allocentric direction, egocentric direction, and spatial location (Hunsaker, 2013; Kesner, 2013).

A temporal (time) attribute within this framework involves memory representations of the duration of a stimulus and the succession or temporal order of temporally separated events or stimuli, and from a time perspective, the memory representation of the past (Kesner and Hunsaker, 2010).

A response attribute within this framework involves memory representations based on feedback from motor responses (often based on proprioceptive and vestibular cues) that occur in specific situations as well as memory representations of stimulus–response associations.

A reward value (affect) attribute within this framework involves memory representations of reward value, positive or negative emotional experiences, and the associations between stimuli and rewards.

A sensory/perceptual attribute within this framework involves memory representations of a set of sensory stimuli that are organized in the form of cues as part of a specific experience. Each sensory modality (olfaction, auditory, vision, somatosensory, and taste) can be considered part of the sensory/perceptual attribute component of memory.

Processes Associated with Each Attribute

Within each system attribute, information is processed in different ways based on different operational characteristics (Table 6.1).

For the event-based memory system, specific processes involve (1) selective filtering or attenuation of interference of temporary memory representations of new information and is labeled pattern separation, (2) encoding of new information, (3) short- and intermediate-term memory for new information, (4) the establishment of

Table 6.2 Primary Neuroanatomical Correlates Underlying Each Attribute in Rodents

Attribute	Event Based	Knowledge Based	Rule Based
Spatial	Hippocampus	Parietal cortex	Infralimbic/prelimbic[a]
			Retrosplenial cortex
Temporal	Hippocampus	Anterior cingulate	Anterior cingulate
	Basal ganglia	Infralimbic/prelimbic[a]	Infralimbic/prelimbic[a]
Sensory/perceptual	Sensory cortices	TE2 cortex[b]	Infralimbic/prelimbic[a]
		Perirhinal cortex	
		Piriform cortex	
Response	Caudoputamen	Precentral cortex	Precentral cortex
		Cerebellum	Cerebellum
Affect	Amygdala	Agranular insula[c]	Agranular insula[c]
			Infralimbic/prelimbic[a]
Executive function	Basal ganglia	Infralimbic/prelimbic[a]	Infralimbic/prelimbic[a]
		Parietal cortex	Parietal cortex
Social		Unknown networks	
Proto-language			

Murine homologs of:
[a]medial prefrontal cortex.
[b]inferior temporal cortex.
[c]orbitofrontal cortex.

arbitrary associations, (5) consolidation or elaborative rehearsal of new information, and (6) retrieval of new information based on flexibility, action, and pattern completion.

For the knowledge-based memory system, specific processes include (1) encoding of repeated information, (2) selective attention and selective filtering associated with permanent memory representations of familiar information, (3) perceptual memory, (4) consolidation and long-term memory storage partly based on arbitrary and/or pattern associations, and (5) retrieval of familiar information based on flexibility and action.

For the rule-based memory system, it is assumed that information is processed through the integration of information from the event- and knowledge-based memory systems for the use of major processes that include (1) the selection of strategies and rules for maintaining or manipulating information for subsequent decision making and action, (2) short-term or working memory for new and familiar information, (3) development of goals, (4) prospective coding, (5) affecting decision processes, and (6) comparing actions with expected outcomes.

Attributes Map onto Neural Substrates

On a neurobiological level each attribute maps onto a set of neural regions and their interconnected neural circuits (Table 6.2). For example, within the event-based memory system, it has been demonstrated that in animals and humans the hippocampus supports memory for spatial, temporal, and language attribute information; the caudate mediates memory for response attribute information; the amygdala subserves memory for reward value (affect) attribute information; and the perirhinal and extrastriate visual cortex support memory for visual object sensory/perceptual attribute.

Within the knowledge-based memory system, it has been demonstrated that in animals and humans the posterior parietal cortex supports memory for spatial attributes; the dorsal and dorsolateral prefrontal cortex and/or anterior cingulate support memory for temporal attributes; the premotor, supplementary motor, and cerebellum in monkeys and humans and precentral cortex and cerebellum in rats support memory for response attributes; the orbital prefrontal cortex supports memory for reward value (affect) attributes; the inferotemporal cortex in monkeys and humans and TE2 cortex in rats subserves memory for sensory/perceptual attributes (e.g., visual objects); and the parietal cortex, Broca and Wernicke's areas subserve memory for the language attribute.

Within the rule-based memory system, it can be shown that different subdivisions of the prefrontal cortex (and rodent homologs; Preuss, 1995; Rose and Woolsey, 1948a,b; Uylings et al., 2003) support different attributes. For example, the dorsolateral and ventrolateral prefrontal cortex in humans support spatial, object, and language attributes and the infralimbic and prelimbic cortex in rats supports spatial and visual object attributes; the premotor and supplementary motor cortex in monkeys and humans and precentral cortex in rats support response attributes; the dorsal, dorsolateral, and mid-dorsolateral prefrontal cortex in monkeys and humans and anterior cingulate in rats mediate primarily temporal attributes; and the orbital prefrontal cortex in monkeys and humans and agranular insular cortex in rats support affect attributes (Kesner, 2000).

Interactions Among Attributes

Despite the relative independence and parallel processing of the different attributes, it bears to mention that the neural systems that underlie memory of all forms interact and contain similar nodes (i.e., hippocampus processes space and time and in special cases can process sensory/perceptual information and affect—but all these attributes are processed in larger networks made up of disparate elements). To provide a concrete example for interactions among attributes, the interaction among temporal, spatial, and sensory/perceptual attributes will be discussed based on different interactions on task demands.

The nature of the interactions between memory systems can be evaluated to dissect out the processes involved in both episodic and nonepisodic behavioral experiments. For illustration, two hippocampus-dependent tasks involving specific and easily identifiable sensory/perceptual stimuli (what), spatial information (where—computed from a combination of sensory/perceptual and temporal attributes), and temporal relationships between the stimuli (when) will be compared and contrasted. One task will require event-based memory processes and the other task can be solved via knowledge-based memory processes. The nature of the interactions between these three attributes corresponding to what, when, and where will be analyzed to differentiate between the two tasks.

The knowledge-based memory task requires that a pair of associations be acquired over multiple training trials. It is an object—trace—place paired-associate task involving sensory/perceptual stimuli (what), a temporal stimulus in the form of a temporal discontinuity (trace interval; when), and spatial information (where). This task is designed as follows: when a particular spatial location (a) and a Garfield toy (1) are paired across a 10-s trace interval (an underscore), the animal is rewarded (a_1+). Also if a different spatial location (b) and a truck toy are matched (2), the animal is rewarded (b_2+). If spatial location (a) and the toy truck are paired, there is no reward (a_2-), similarly for spatial location (b) and (a) Garfield toy (b_1-). The trace interval separates the sensory/perceptual stimulus and the presentation of the spatial location (Hunsaker et al., 2006). If the association presented during a trial were rewarded, the rat would receive a reward upon displacing a block in the correct spatial location, which is then represented by the affect attribute, signaling a correct choice. This should bind the sensory/perceptual stimulus and spatial location association across the temporal discontinuity (an association involving what, when, and where). Then, the animal is presented with a new sensory/perceptual stimulus and spatial location association. If rewarded, then the process continues as before; if not rewarded, the animal does not receive any reward, and the affect attribute signals an error. Learning this task within only the event-based memory system would be difficult because the event-based memory system is susceptible to trial-by-trial interference. Both temporally adjacent (e.g., subsequent) and spatially adjacent (e.g., occurring in the same or very similar spatial locations irrespective of temporal contiguity) episodes would interfere and degrade each other during acquisition.

Learning these associations involves comparing accumulated behavioral episodes or events within the knowledge- and rule-based memory systems to develop appropriate rules, goals, and schemas to perform the task efficiently. Also, these two latter systems generate and apply abstract rules and generalize temporal, spatial, and internal contexts. In other words, the knowledge- and rule-based systems read the accumulated behavioral episodes, clarify the relevant contextual information, and apply this information to guide future actions. Once the knowledge- and rule-based memory systems have processed the data and generated the schemas necessary to perform the task, the event-based memory system does not significantly contribute to performance of this task since the four discriminations or associations (a_1+, b_2+, a_2-, b_1-) have been efficiently encoded and only need to be discriminated from each other (O'Reilly and Frank, 2006; O'Reilly and Rudy, 2001).

In contrast to the above biconditional discrimination, a task developed by Day et al. (2003) and modified by Kesner and colleagues (2008) allows rats to perform a very similar sensory/perceptual stimulus and spatial location association in an event-based manner. During the study phase, the animal receives two rewarded object—place pairings (i.e., single sensory/perceptual stimulus in a spatial location defined by the sum total of sensory/perceptual stimuli in the environment) separated by a short temporal interval. Since there are two distinct behavioral episodes in close temporal proximity to each other, information pertaining to temporal relationships between stimuli (e.g., temporal contiguity) discriminates the two episodes and facilitates retrieval (Hunsaker and Kesner, 2008). During the test phase, the animal is provided with a retrieval cue. The animal has to learn that the sensory/perceptual stimulus provided as a retrieval cue is a signal to displace a neutral block in the corresponding spatial location previously paired with the cue (or to the sensory/perceptual stimulus cued by a spatial location). Since none of the 50 sensory/perceptual stimuli and 48 spatial locations are frequently paired (there are nearly 2500 possible combinations), each pairing is trial (or behavioral episode) unique. Since the animal receives two distinct behavioral episodes followed by a retrieval cue to signal which of the two episodes needs to be recalled, the animal has not only to remember the relevant episode to receive reward, but also to discriminate between the relevant episode and the episodes presented either immediately before or after the relevant episode, as well as all previous episodes that occurred in the same or a similar spatial context.

The critical difference between the two tasks is not the cued-recall nature of the latter task *per se* but that the associations to be remembered are trial unique. This allows each behavioral episode to be coded as unique but increases potential interference from previous or subsequent behavioral episodes. To overcome this interference and to guide efficient recall of the correct behavioral episode, this task is performed with the contribution of the knowledge- and rule-based systems such as traditional biconditional discrimination tasks, but the trial specific episodes make it necessary to depend on the event-based memory system to compare the retrieval cue to the stored episodes to efficiently recall the correct, and only the correct, behavioral episode to guide behavioral decisions and actions.

Applying the Attribute Model

General Advice

Despite the need to move beyond limiting behavioral research to the standard behavioral paradigms (i.e., water maze, contextual fear conditioning), it is by no means necessary to avoid these tasks all together. Rather, it is important to integrate these tasks more through behavioral analysis necessary for elucidating behavioral endophenotypes (Table 6.3 for lists of tasks used for general phenotyping as well as behavioral endophenotyping; Hunsaker, 2012a,b for references to the specific tasks).

To begin, it is important to evaluate the basic sensory function in all mice, because any deficits in basic sensation or perception confound interpretations of behavioral results (Crawley, 2007). Sensory deficits do not preclude the behavioral analysis of a mouse model. When a mouse shows sensory deficits, either the model can be bred onto a different background strain over numerous generations—commonly >10 generations backcrossed onto the C57BL/6J strain—or else behavioral tasks can be

Table 6.3 Summary of Behavioral Tasks Commonly Used in Behavioral Phenotyping
Strategies Organized by General Domain

Attribute Tested	Behavioral Phenotyping	Behavioral Endophenotyping
Memory (spatial, temporal)	Water maze Radial arm maze Barnes maze Active/passive Avoidance Contextual fear conditioning	
Spatial memory		Categorical (Metric) Processing Coordinate (Topological) Processing Touchscreen pattern Separation Delay match to place with variable interference Delay match to place with variable cues
Temporal memory		Trace fear conditioning Temporal ordering of stimuli Sequence learning tasks Sequence completion tasks Duration discrimination
Associative		Biconditional discrimination Cued-recall task for trial unique associations
Affect	Classical fear conditioning Open field Elevated plus maze Porsolt test	
Valence		Reward contrast with variable Reward value
Anhedonia		Anticipatory contrast task Species relevant sexual behaviors
Approach-avoidance		Hyponeophagia Defensive burial
Fear processing		Defensive test battery Classical, contextual, trace fear conditioning
Motor (response)	Rotarod	
Visuomotor		Skilled forelimb reaching Capellini handling task Seed shelling tasks Parallel beam or ladder walking tasks

(Continued)

Table 6.3 (Continued)

Attribute Tested	Behavioral Phenotyping	Behavioral Endophenotyping
Motor learning		Acquisition of Skilled Reaching
		Acquisition of Rotarod (initial training)
		Working Memory for Motor Movements
Sensory/perceptual	Prepulse inhibition	Prepulse inhibition
	Acoustic startle	Acoustic startle
	Hot plate analgesia	Psychonomic threshold
Social	Three chamber social novelty	Social dyadic behavior
		Resident intruder tests
		Social transmission of food preference
		Social dominance
Executive function	Operant conditioning	
	Holeboard exploration	
	Reversal learning	
Cognitive control		Contextually cued biconditional discrimination
		Serial reversal learning
		Stop signal task
		Probabalistic (80/20) reversal learning
Attention		5 Choice serial reaction time task
		Covert attention tasks

Also summarized are behavioral tasks proposed to be useful for behavioral endophenotyping organized by component attributes (Hunsaker, 2012a,b for references for each task that emphasize the methods for each paradigm).

chosen that minimize the contribution of the particular sensory modality that is not being processed, as Farley and coworkers (2011) demonstrated in an experiment emphasizing behaviors to test space and temporal processing while concurrently not requiring vision in mice that are blind from an early age.

After evaluating basic sensory function in the mouse model, it is critical to determine the pattern of behavioral strengths and weaknesses in the population being modeled by the mouse. With this information from the clinical population, it is important to either create or adopt behavioral tasks to evaluate the same cognitive attributes or domains as tested in the clinical population. For example, if a disorder being modeled shows global memory deficits (measured by intelligence (IQ) and neuropsychological tests) without concomitant impairments for executive function, then the mouse model needs to be tested for memory across a number of domains or attributes evaluated by the neuropsychological tests in order to better dissect cognitive function in the model. In this example, executive function should also be evaluated, but in this case to verify intact executive function in the model.

More concretely, a general memory deficit may be mediated by an inability to encode new information, consolidate/retrieve encoded information, or understand the rules required to perform correctly on a given test. All of these factors can be tested in mice and can further be evaluated across domains: spatial, temporal, and response memory can be specifically evaluated in the mouse, as can the contribution of affect to memory, anxiety, and depressive behaviors. With these data, research into the mouse model may actually serve to inform the clinic as to more specific domains that can be tested in the clinic—emphasizing a direct interaction across the research in the clinic and the behavioral genetics laboratory (Hunsaker, 2012a,b; Table 6.3).

There are a number of difficult questions that must be answered in the development of a comprehensive behavioral endophenotyping approach, including: How many of these behavioral tests could be conducted on a single cohort of mice? Would multiple tests on the tasks listed lead to potential confounds? What are the potential contributions/confounds of environmental effects, such as housing conditions, on mouse behavior? These are important factors, because small changes in experimental design can make the difference between a successful analysis of a behavioral phenotype and a collection of uninterpretable data.

The question of testing the same group or mice across multiple experiments is sometimes more complicated as it seems, but a solution can be designed if viewed through the attribute model. The most critical aspects that need to be taken into account when performing multiple experiments on the same mice are twofold: any role for negative effect on task performance, and the rule-based memory system succumbing to interference across tasks. It is important to keep in mind the role for negative effect for task performance not only on the present task, but also on the future tasks performed with the same group of mice. If a mouse is to receive fear conditioning in the middle of an experimental design, it will take a week or so of handling for the mouse to unlearn any associations between the fear conditioning and the experimenter (Rudy and O'Reilly, 2001).

For the rule-based memory system, it is important to remember that mice and rats take a significantly longer amount of time to learn and apply rules to guide behavior than humans. As such, if a researcher wants to perform any experiments that require the mouse to learn a rule or set of rules to guide behavior, and are not just exploiting the natural tendency of mice to explore their environment and behaviorally respond to novelty across domains (i.e., novelty detection), then intervening tasks not requiring rule learning/implementation need to be presented prior to the usage of another task taxing the rule-based memory system. Also important in this case is the use of very different apparatus for each rule-based learning task, or else previously learned rules will have to be explicitly extinguished prior to beginning training on a new task (Cohen and O'Reilly, 1996).

As a rule of thumb, it is best to perform at most a set of experiments evaluating basic sensory function in mice (Crawley, 2007), followed by tasks evaluating each attribute in turn, followed potentially by the water maze/avoidance tasks and fear conditioning as the final task to prevent carryover from experiments interfering with subsequent experiments. In this, most likely, more than one experiment tasking executive function/complex rule learning can be performed. More than one of these tasks will result in interference that will confound interpretation of subsequent rule-based tasks. Additionally, aside from the water maze and fear conditioning experiments, it is recommended that series of experiments be counterbalanced across animals and groups, preferably using a Latin square design to reduce the contribution of task order to any observed effects. Additionally, it is often worth testing the ability of mice to perform the behavioral task battery by evaluating a few wild-type mice of the background strain being used to verify that they can perform all the tasks without being overwhelmed by excessive testing.

In addition to genetic background, the treatment of mice prior to and during experimentation is critical. It has been shown a number of times that alterations to the cage environment is an important factor for later behavioral improvement—such that an enriched environment results in better performance on behavioral tasks and increased gray matter and dendritic complexity. As such, mice should preferably be housed in a standard fashion, either with a set number of mice per cage or else singly housed, but it is important to note that mice do not do as well singly housed as rats; they tend to show increased anxiety levels, which may affect task performance (Van de Weerd et al., 2002). As such, multiple housed is recommended unless precise drug dosing or food deprivation is required that precludes group housing. Furthermore, the amount of stimuli available to the mice within each cage should be uniform across cages. As such, a standard environment needs to be maintained among all mice during experimentation.

Specific Application of the Attribute Model

To properly apply the attribute model to optimally select behavioral paradigms for the study of genetic models, a researcher must ask themselves a number of questions (in no particular order). (1) What are the pathological effects of the mutation on anatomy? In other words, in what way is the structure of brain regions, neuroendocrine regions, receptors, etc. altered by the mutation. The answer to this question suggests a starting point for task selection.

For considering the pathological effects of mutations on anatomy, the most efficient way to specifically quantify this effect is in collaboration with a veterinarian or (neuro)anatomist that studies the species. A full histological profile can be performed by core facilities at most universities and the necessary sectioning and staining techniques are easily trained to skillful technicians. Depending upon the hypotheses being evaluated, it may be helpful to evaluate the organ systems of the mutant rat or mouse. A key example from my own work has demonstrated that key neuropathologic features of a mouse model for the fragile X premutation (the CGG KI mouse) were also present in the peripheral nervous system as well as somatic organs and cardiac muscle—similar to that observed in parallel in postmortem tissue from human premutation carriers (Hunsaker et al., 2011a,b). Unfortunately, it never occurred to any of the researchers involved to look at these features in mice years before we finally did so. Such information would have been able to guide human research into the pathology of the fragile X premutation years before this report. Using the information provided in Table 6.3 along with a more thorough literature search would inform a first wave of behavioral tasks to consider for evaluating the behavioral consequences of the genetic mutation.

Additionally, we were able to identify pathology associated with the premutation throughout the brain, but specific patterns of development were identified that led us to evaluating hippocampus and parietal cortex function in the CGG KI mice (Hunsaker et al., 2009; Schluter et al., 2012; Wenzel et al., 2010).

(2) Is there anything known from human research concerning the effects of mutations within the genetic locus being evaluated? If so, then this provides a starting point, that of generating a behavioral model in a mouse or rat for the research that has been undertaken in the clinic. Upon replicating (or more to the point recapitulating) these effects, logical extensions to better characterize the deficits can be undertaken.

This is a generally self-explanatory process, but nonetheless it bears extrapolation. Often, there are subclinical effects for a number of populations with genetic mutations that are important for animal researchers to know as a mouse model may be able to tease out these subclinical effects in the nonhuman models in ways not possible to do in patients. What this means is that, in addition to reading the literature available on a given mutation or disease, it is often much more efficient to initiate a collaboration, or even just initiate contact with a clinician and probe their experiences and ask their impressions on what is worth studying in the population.

For a specific example from my own work in the CGG KI mouse model of the fragile X premutation, the process by which experimental hypotheses arose will be discussed. The full clinical manifestation of the fragile X premutation is a late onset neurodegenerative disorder called fragile X-associated tremor/ataxia syndrome, or FXTAS. FXTAS is primarily a movement disorder characterized by Parkinsonism, intention/kinetic tremor, and cerebellar gait ataxia, along with a dysexecutive syndrome and cognitive decline leading to dementia. In the initial screen of the mouse, no convincing phenotypes were found to model FXTAS (van Dam et al., 2005). Upon discussion with collaborators studying the clinical population, it was brought to our attention that the population constantly report "muddy thinking," and what can best be described by general clumsiness and subclinical apraxia (reports of tripping over their own feet, not being athletic, or spilling their milk at lunch).

Based upon these verbal reports and models being developed to account for these reports, it was determined that evaluating spatial and temporal processing was an appropriate starting point and evaluating motor tasks was worth a try in the mouse model, so long as the tasks were sufficiently difficult. Though not reviewed here, these experiments were able to identify early onset, progressive deficits for spatiotemporal memory as well as motor difficulty on difficult, but not relatively easy, motor tests (Borthwell et al., 2012; Diep et al., 2012; Hunsaker et al., 2010, 2012, 2011, 2009). In other words, the clinical manifestation of FXTAS was not recapitulated in the mouse model, but the subclinical manifestation of the fragile X premutation observed by our clinical collaborators were recapitulated.

(3) What molecular cascades are affected by the mutation? The answer to this question will set up research questions into whether the molecular cascade disrupts more short-term learning, or consolidation, or long-term retrieval processed preferentially.

One way of thinking about this is to consider that there are mutations that affect induction of long-term potentiation or else affect early-phase Long Term Potentiation (LTP) separately from mutations that preferentially affect late-phase Long Term Potentiation (LTP). As memory processes can be similarly parceled, the timescale of the mutations effect on behavior be considered. Again with an example from my own research, it is important to assess what precisely is the molecular pathology associated with the fragile X premutation, since the same parametric mutation underlies both premutation ([50,200] CGG repeats) and full mutation (230 + CGG repeats) underlying fragile X syndrome. It has been demonstrated in the Fmr1 KO mouse model of fragile X syndrome that the lack of Fmrp results in an exaggerated long-term depression dependent upon Group 1 mGluR receptors (mGlur1/5; both paired pulse low-frequency stimulation in the presence of D-(-)-2-Amino-5-phosphonopentanoic acid (APV) APV as well as bath application of (S)-3,5-Dihydroxyphenylglycine (DHPG) DHPG; Huber et al., 2002). Since the fragile X premutation and the CGG KI mice show reduced, but nonzero, Fmrp, it was unknown whether the characteristic findings of enhanced mGluR1/5-dependent Long Term Depression (LTD) would be present in the CGG KI mice. What was found was that, in fact, the CGG KI mouse showed a CGG repeat length-dependent impairment in plasticity induction for both NMDA LTP, LTD, and mGluR1/5 LTD, but no enduring effects after plasticity was induced. We interpreted these findings as suggestive of more general cellular dysfunction not particularly dependent upon one receptor system over another (Hunsaker et al., 2012).

Intriguingly, an independent group demonstrated in a different mouse model of the fragile X premutation (CGG-CCG mouse) that shows much more profound reductions in Fmrp levels phenocopied the Fmr1 KO mouse for mGluR1/5 LTD. They did not look at NMDA-receptor dependent plasticity (Iliff et al., 2013). These findings, although they appear discrepant, actually make perfect sense based upon the hypothesis that reduced Fmrp levels results in exuberant mGluR1/5 activity; and as such both support theories of premutation-dependent molecular pathologies.

As can be seen from the specific examples provided earlier, there are a number of steps to choose a behavioral paradigm to test a rodent model. Importantly, although presented earlier as a holistic approach, it is possible to ask the questions serially based on the laboratory expertise. If a laboratory has a particular focus on molecular mechanisms, then obviously the elucidation of biochemical pathways affected by the mutation are the logical starting point. Similar logic applies to cognitive neuroscientists and anatomists in approaching research questions.

Conclusions

An important consideration in the study of mouse models of any genetic disease is how well the behavior of the mouse serves to actually model the behaviors present in the clinical population being modeled. In order to optimally address this concern is to explicitly develop a behavioral endophenotype of the mouse model in which the mouse is explicitly tested using tasks rigorously designed to explicitly test the cognitive domains affected in the clinical population. To achieve this goal, it is important not only to apply standard behavioral tasks to mouse models of genetic disorders, but also to directly evaluate brain function across all cognitive domains. Furthermore, it is critical for behavioral genetics laboratories to interact with clinical research laboratories to develop comprehensive behavioral endophenotypes for the disorder being modeled. The strength of behavioral genetics and the mouse models is the ability to apply behavioral paradigms known to be subserved by known anatomical loci not only to determine the behavioral phenotype of the model, but also to elucidate candidate brain regions affected by the mutation (Robbins et al., 2012). When research into mouse models of genetic disorders emphasizes patterns of mnemonic strengths and weaknesses across domains (i.e., the behavioral endophenotype), the results will not only directly model the disorder being studied, but also serve both as risk prodrome for disease onset or progression and as outcome measures that can be applied by the clinic in interventional studies.

In addition to this interaction, it is rather difficult to elucidate a behavioral endophenotype using only a limited toolkit. As such, the attribute model is unique in that the theory itself was designed to not only explain and describe processes underlying memory, but also provide hypotheses that can specifically be tested. As mentioned earlier, Table 6.3 contains a list of the behavioral tasks available to the researcher today that have been validated in mice and rats to test specific memory processes (Hunsaker, 2012a,b for references corresponding to the tasks). Using subsets of these tasks in turn with collaborations with molecular and clinical neuroscientists we were able to design a battery of test to evaluate a behavioral endophenotype in the CGG KI mouse. Importantly, research into the CGG KI mouse has critically informed the clinic and provided rationale to formally identify subclinical apraxia and postural issues early in life as well as emphasized the need to evaluate spatiotemporal processing in the fragile X premutation (Goodrich-Hunsaker et al., 2011a,b,c; Narcisa et al., 2011; Wong et al., 2012).

In broader application, it is quite possible that applying models such as the attribute processing model concurrent with translational interactions among molecular, behavioral, and clinical researchers may result in a paradigm shift in behavioral genetics. At present the field is working toward developing standardized sets of behavioral cores that can process large numbers of genetic models in relatively short times. This is commendable, but if the research culture opens up to free exchange of theories and hypotheses between molecular genetics and clinical neuroscience laboratories as well as general psychiatry and medical research into clinical populations, then the mouse model will be a tool to identify neurocognitive as well as molecular endpoints of mutations states. Such a tool is begging to be developed to facilitate the testing of candidate compounds prior to phase I clinical trials to reduce the tendency of compounds to succeed in the mouse and fail in the human. Additionally, such tools also would facilitate the elucidation of what the behavioral consequences of specific alterations to biochemical processes actually are at a much higher resolution than has been possible to date.

References

Borthwell, R.M., Hunsaker, M.R., Willemsen, R., Berman, R.F., 2012. Spatiotemporal processing deficits in female CGG KI mice modeling the fragile X premutation. Behav. Brain Res. 233 (1), 29–34.

Churchwell, J.C., Kesner, R.P., 2011. Hippocampal-prefrontal dynamics in spatial working memory: interactions and independent parallel processing. Behav. Brain Res. 225 (2), 389–395.

Churchwell, J.C., Morris, A.M., Heurtelou, N.M., Kesner, R.P., 2009. Interactions between the prefrontal cortex and amygdala during delay discounting and reversal. Behav. Neurosci. 123 (6), 1185–1196.

Churchwell, J.C., Morris, A.M., Musso, N.D., Kesner, R.P., 2010. Prefrontal and hippocampal contributions to encoding and retrieval of spatial memory. Neurobiol. Learn. Mem. 93 (3), 415–421.

Cohen, J.D., O'Reilly, R.C., 1996. A preliminary theory of the interactions between prefrontal cortex and hippocampus that contribute to planning and prospective memory. In: Brandimonte, M., Einstein, G.O., McDaniel, M.A. (Eds.), Prospective Memory: Theory and Applications. Lawrence Erlbaum Associates, Mahwah, NJ, pp. 267–295.

Crawley, J.N., 2007. Mouse behavioral assays relevant to the symptoms of autism. Brain Pathol. 17, 448–459.

Day, M., Langston, R., Morris, R.G.M., 2003. Glutamate-receptor-mediated encoding and retrieval of paired-associate learning. Nature. 424, 205–209.

Diep, A.A., Hunsaker, M.R., Kwock, R., Kim, K., Willemsen, R., Berman, R.F., 2012. Female CGG knock-in mice modeling the fragile X premutation are impaired on a skilled forelimb reaching task. Neurobiol. Learn. Mem. 97 (2), 229–234.

Farley, S.J., McKay, B.M., Disterhoft, J.F., Weiss, C., 2011. Reevaluating hippocampus dependent learning in FVB/N mice. Behav. Neurosci. 125, 871–878.

Goodrich-Hunsaker, N.J., Wong, L.M., McLennan, Y., Srivastava, S., Tassone, F., Harvey, D., et al., 2011a. Young adult female fragile X premutation carriers show age- and genetically-modulated cognitive impairments. Brain Cogn. 75 (3), 255–260.

Goodrich-Hunsaker, N.J., Wong, L.M., McLennan, Y., Tassone, F., Harvey, D., Rivera, S.M., et al., 2011b. Enhanced manual and oral motor reaction time in young adult female fragile X premutation carriers. J. Int. Neuropsychol. Soc. 17 (4), 746–750.

Goodrich-Hunsaker, N.J., Wong, L.M., McLennan, Y., Tassone, F., Harvey, D., Rivera, S.M., et al., 2011c. Adult female fragile X premutation carriers exhibit age- and CGG repeat length-related impairments on an attentionally based enumeration task. Front. Hum. Neurosci. 5, 63.

Huber, K.M., Gallagher, S., Warren, S.T., Bear, M.F., 2002. Altered synaptic plasticity in a mouse model of fragile X mental retardation. Proc. Natl. Acad. Sci. USA. 99 (11), 7746–7750.

Hunsaker, M.R., 2012a. The importance of considering all attributes of memory in behavioral endophenotyping of mouse models of genetic disease. Behav. Neurosci. 126 (3), 371–380.

Hunsaker, M.R., 2012b. Comprehensive neurocognitive endophenotyping strategies for mouse models of genetic disorders. Prog. Neurobiol. 96, 220–241.

Hunsaker, M.R., 2013. Embracing complexity: using the attribute model to elucidate the role for distributed neural networks underlying spatial memory processes. OA Neurosciences. 1 (1), 2.

Hunsaker, M.R., Kesner, R.P., 2008. The attributes of episodic memory processing. In: Dere, E., Huston, J.P., Easton, A., Nadel, L. (Eds.), Handbook of Behavioral Neuroscience, Vol. 18, Episodic Memory Research. Elsevier, Amsterdam, pp. 57−79.

Hunsaker, M.R., Kesner, R.P., 2013. The operation of pattern separation and pattern completion processes associated with different attributes or domains of memory. Neurosci. Biobehav. Rev. 37 (1), 36−58.

Hunsaker, M.R., Thorup, J.A., Welch, T., Kesner, R.P., 2006. The role of CA3 and CA1 in the acquisition of an object−trace−place paired associate task. Behav. Neurosci. 120, 1252−1256.

Hunsaker, M.R., Wenzel, H.J., Willemsen, R., Berman, R.F., 2009. Progressive spatial processing deficits in a mouse model of the fragile X premutation. Behav. Neurosci. 123 (6), 1315−1324.

Hunsaker, M.R., Goodrich-Hunsaker, N.J., Willemsen, R., Berman, R.F., 2010. Temporal ordering deficits in female CGG KI mice heterozygous for the fragile X premutation. Behav. Brain Res. 213 (2), 263−268.

Hunsaker, M.R., Greco, C.M., Spath, M.A., Smits, A.P., Navarro, C.S., Tassone, F., et al., 2011a. Widespread non-central nervous system organ pathology in fragile X premutation carriers with fragile X-associated tremor/ataxia syndrome and CGG knock-in mice. Acta Neuropathol. 122 (4), 467−479.

Hunsaker, M.R., von Leden, R.E., Ta, B.T., Goodrich-Hunsaker, N.J., Arque, G., Kim, K., et al., 2011b. Motor deficits on a ladder rung task in male and female adolescent and adult CGG knock-in mice. Behav. Brain Res. 222 (1), 117−121.

Hunsaker, M.R., Kim, K., Willemsen, R., Berman, R.F., 2012. CGG trinucleotide repeat length modulates neural plasticity and spatiotemporal processing in a mouse model of the fragile X premutation. Hippocampus. 22 (12), 2260−2275.

Iliff, A.J., Renoux, A.J., Krans, A., Usdin, K., Sutton, M.A., Todd, P.K., 2013. Impaired activity-dependent FMRP translation and enhanced mGluR-dependent LTD in fragile X premutation mice. Hum. Mol. Genet. 22 (6), 1180−1192.

Kesner, R.P., 2000. Subregional analysis of mnemonic functions of the prefrontal cortex in the rat. Psychobiology. 28, 219−228.

Kesner, R.P., 2013. Neurobiological foundations of an attribute model of memory. Compar. Cogn. Behav. Rev. 8, 29−59.

Kesner, R.P., Churchwell, J.C., 2011. An analysis of rat prefrontal cortex in mediating executive function. Neurobiol. Learn. Mem. 96 (3), 417−431.

Kesner, R.P., Hunsaker, M.R., 2010. Temporal attributes of episodic memory. Behav. Brain. Res. 215 (2), 299−309.

Kesner, R.P., Hunsaker, M.R., Warthen, M.W., 2008. The CA3 subregion of the hippocampus is critical for episodic memory processing by means of relational encoding in rats. Behav. Neurosci. 122 (6), 1217−1225.

Narcisa, V., Aguilar, D., Nguyen, D.V., Campos, L., Brodovsky, J., White, S., et al., 2011. Quantitative assessment of tremor and ataxia in female FMR1 premutation carriers using CATSYS. Curr. Gerontol. Geriatr. Res. 2011, 484713.

O'Reilly, R.C., Frank, M.J., 2006. Making working memory work: a computational model of learning in the prefrontal cortex and basal ganglia. Neural Comput. 18, 283−328.

O'Reilly, R.C., Rudy, J.W., 2001. Conjunctive representations in learning and memory: principles of cortical and hippocampal function. Psychol. Rev. 108, 311−345.

Preuss, T.M., 1995. Do rats have prefrontal cortex? The Rose−WoolseyAkert program reconsidered. J. Cogn. Neurosci. 7 (1), 1−24.

Robbins, T.W., Gillan, C.M., Smith, D.G., de Wit, S., Ersche, K.D., 2012. Neurocognitive endophenotypes of impulsivity and compulsivity: towards dimensional psychiatry. Trends Cogn. Sci. 16, 81−91.

Rose, J.E., Woolsey, C.N., 1948a. The orbitofrontal cortex and its connections with the mediodorsal nucleus in rabbit, sheep and cat. Res. Publ. Assoc. Res. Nerv. Ment. Dis. 27 (1), 210−232.

Rose, J.E., Woolsey, C.N., 1948b. Structure and relations of limbic cortex and anterior thalamic nuclei in rabbit and cat. J. Comp. Neurol. 89 (3), 279−347.

Rudy, J.W., O'Reilly, R.C., 2001. Conjunctive representations, the hippocampus, and contextual fear conditioning. Cogn. Affect Behav. Neurosci. 1, 66−82.

Schluter, E.W., Hunsaker, M.R., Greco, C.M., Willemsen, R., Berman, R.F., 2012. Distribution and frequency of intranuclear inclusions in female CGG KI mice modeling the fragile X premutation. Brain Res. 1472, 124−137.

Squire, L.R., 1994. Declarative and nondeclarative memory: multiple brain systems supporting learning and memory. In: Schacter, D.L., Tulving, E. (Eds.), Memory Systems. MIT Press, Cambridge, pp. 203−231.

Tulving, E., 1983. Elements of Episodic Memory. Clarendon Press, Oxford.

Uylings, H.B., Groenewegen, H.J., Kolb, B., 2003. Do rats have a prefrontal cortex? Behav. Brain Res. 146 (1-2), 3−17.

Van Dam, D., Errijgers, V., Kooy, R.F., Willemsen, R., Mientjes, E., Oostra, B.A., et al., 2005. Cognitive decline, neuromotor and behavioural disturbances in a mouse model for fragile-X-associated tremor/ataxia syndrome (FXTAS). Behav. Brain Res. 162 (2), 233−239.

Van de Weerd, H.A., Aarsen, E.L., Mulder, A., Kruitwagen, C.L.J.J., Hendricksen, C., Baumans, V., 2002. Effects of environmental enrichment for mice: variation in experimental results. J. Appl. Anim. Welf. Sci. 5, 87−109.

Wenzel, H.J., Hunsaker, M.R., Greco, C.M., Willemsen, R., Berman, R.F., 2010. Ubiquitin-positive intranuclear inclusions in neuronal and glial cells in a mouse model of the fragile X premutation. Brain Res. 1318, 155−166.

Wong, L.M., Goodrich-Hunsaker, N.J., McLennan, Y., Tassone, F., Harvey, D., Rivera, S.M., et al., 2012. Young adult male carriers of the fragile X premutation exhibit genetically modulated impairments in visuospatial tasks controlled for psychomotor speed. J. Neurodev. Disord. 4 (1), 26.

7 Protein Synthesis and Memory: A Word of Caution

Roberto Agustín Prado-Alcalá, Andrea C. Medina, Norma Serafín and Gina L. Quirarte

Departamento de Neurobiología Conductual y Cognitiva, Instituto de Neurobiología, Universidad Nacional Autónoma de México, Campus Juriquilla, Querétaro, Qro 76230, México

A little over a century ago Georg Elias Müller and his disciple Alfons Pilzecker published a monograph where they introduced the concept of "consolidation" into the field of Experimental Psychology. They defined it as a time-dependent process that allows the transference of learned information from a short-term memory store to a long-term memory store, thus implying that the formation of long-lasting memory is not instantaneous. Their experiments also led them to conclude that memories are fragile during the period of consolidation (Müller and Pilzecker, 1900).

These pioneering ideas have guided, directly or indirectly, all of the research related to the neurobiology of memory. A myriad of experiments have demonstrated that recent memories stabilize over time, but during its formation they are vulnerable to influences that may jeopardize or improve its storage. The consolidation hypothesis is also supported by evidence suggesting that the impairments of memory induced by posttraining treatments are permanent (Chevalier, 1965; Luttges and McGaugh, 1967). The fact that treatments that interfere with cerebral activity, such as electroconvulsive shocks, protein synthesis inhibitors (PSIs), or drugs that negatively modify synaptic functions have no effect, or limited effect, on memory formation when administered several hours after a learning experience, is taken as evidence of memory consolidation (Lechner et al., 1999; McGaugh, 1966, 2000).

While the experimental evidence for the existence of a consolidation process of memory was multiplying, another, just as important, line of research was evolving. It had to do with the question: What is the physical substrate of the memory store? Even before the postulation of the consolidation hypothesis, the great Spanish neurohistologist Santiago Ramón y Cajal speculated that "cerebral gymnastics" would very probably modify the dendritic architecture as well as axon collaterals in those brain areas that are more related with mental exercising; new associations between sets of neurons would be strengthened by learning. Thus, learning would be expressed in the form of neuronal anatomical changes (Ramon y Cajal, 1899).

Identification of Neural Markers Accompanying Memory. DOI: http://dx.doi.org/10.1016/B978-0-12-408139-0.00007-9

This idea was revived half a century later by Donald Hebb (1949), who talked about the theory of the dual trace of memory formation. Simply stated, this theory proposes that short-term memory is sustained by reverberating activity in local neuronal circuits. This activity may induce structural changes at the synapses of these circuits thus allowing for the permanence of learned information for extended periods of time (long-term memory). This clearly is an updated version of Cajal's speculation.

In 1950, Joseph Katz and Ward Halstead put forward the idea that neuronal proteins may play an important role in the storage of learned information (Katz and Halstead, 1950). The obvious implication of this idea was that through new protein synthesis changes in cerebral citoarchitecture would come about, such as production and growth of neuronal membranes (e.g., in dendrites and spines) and of axonal branches, thought to be needed to improve interneuronal communication. It was not until the late 1950s and early 1960s that antibiotics that inhibited animal cell protein synthesis were described (Ennis and Lubin, 1964; Yarmolinsky and Haba, 1959). All these historic events seemed to be in harmony and set the stage for the experimental testing of the hypothesis that memory consolidation, the alleged process necessary for the formation of long-term memory, is critically dependent upon new protein synthesis in neural tissue.

Thus, in the early 1960s, the laboratory of the Flexners presented the first evidence that PSIs impede the consolidation of memory (Flexner et al., 1962, 1963).

These findings were readily confirmed by Agranoff and Klinger (1964) and by Barondes and Cohen (1966). These pioneering discoveries were followed by a host of papers that gave strong support to the proteininc theory of memory consolidation. For example, it was found, in mice, that systemic administration of PSIs with different mechanisms of action produces significant deficits of long-term memory in a wide variety of learning tasks; the amnesia that was produced is dose-dependent, and it correlates with the degree of PSI (for reviews see Davis and Squire, 1984; Gold, 2008; Martinez et al., 1981; Routtenberg and Rekart, 2005).

The deficiencies in consolidation produced by PSIs have also been reported in birds (Bull et al., 1976; Freeman et al., 1995), fish (Agranoff et al., 1965), moluscs (Agin et al., 2003; Crow and Forrester, 1990), insects (Barraco et al., 1981; DeZazzo and Tully, 1995), and earthworms (Watanabe et al., 2005). These results have been interpreted to mean that some protein synthesis-dependent mechanisms are quite preserved along the phylogenic scale to allow for the formation of memory.

It is currently believed that memory consolidation is dependent, to a large extent, upon cascades of molecular events that induce changes in the physiological and morphological properties of neurons (Alberini, 2009; Atkins et al., 1998; Bailey et al., 1996; Blum et al., 1999; Goelet et al., 1986; Izquierdo et al., 2008; Sweatt, 2001). These events include activation of transcriptional regulators produced by cascades of second messengers, as well as multiple waves of mRNA and protein synthesis.

In March 2013, we run a bibliographic search on PubMed Central (U.S. National Institutes of Health's National Library of Medicine) with the descriptors "protein synthesis AND memory"; there was a turnout of more than 35,000 articles. This alone gives a clear idea about the importance and prevalence of this particular field of the neurobiological study of memory.

In sum, a great deal of experimental evidence indicates that there are two types of memory store (short and long term), that the transfer of information from the short-term store to the long-term store is accomplished through the process of consolidation, and that memory consolidation is dependent upon protein synthesis. The year 2000 marked the hundredth anniversary of the key publication of Müller and Pilzecker (1900) where the phenomenon of consolidation was first described. Up to this point it seemed that the field of the neurobiology of memory was harmoniously and steadily advancing and, all in all, the experimental evidence was congruent with the consolidation hypothesis and with the theory that protein synthesis was a prerequisite for consolidation and, therefore, for long-term memory storage. However, publications that were inconsistent with this theory appeared, as we shall see next.

Under certain experimental conditions, systemic administration of PSIs that produce inhibition of protein synthesis of about 90% for several hours does not impede long-term memory formation (Barondes and Cohen, 1967a; Flexner and Flexner, 1966), or they produce an amnestic state that is followed by recovery of memory when "reminder" stimuli (Quartermain et al., 1970), an ample variety of stimulant drugs (Martinez et al., 1981), and extended training (Barondes and Cohen, 1967b) are given. Also, increased strength of training impedes the amnesic effect of PSIs (Díaz-Trujillo et al., 2009; Flood et al., 1972, 1975).

Other lines of research led to suggest that the amnestic effects of PSIs are due to unspecific effects of such drugs and not to the inhibition of protein synthesis itself. For example, the common PSIs used in this field inhibit the synthesis of catecholamines and increase the stores of tyrosine (Flexner and Goodman, 1975; Flexner et al., 1973; Freedman et al., 1982), and it is known that catecholamines are involved in memory formation (Haycock et al., 1977; Hraschek and Endroczi, 1978).

PSIs also affect nuclear molecules that are implicated in the formation of memory (Bailey et al., 1996; Bevilaqua et al., 2003; Bozon et al., 2003; Cammarota et al., 2000; Guzowski, 2002; Jones et al., 2001; Mazzucchelli et al., 2002), such as the early genes *c-jun*, *c-fos*, and *zif268* (Törocsik and Szeberenyi, 2000). Furthermore, superinduction of early gene mRNA expression by some PSIs leaves open the possibility put forward by Routtenberg and Rekart (2005) that PSIs might exert their amnestic effects through a delayed increment of some protein synthesis and not to inhibition of the synthesis. These authors also proposed that posttranslational modification of proteins already at the synapse is the crucial instructive mechanism underlying long-lasting memory (Routtenberg and Rekart, 2005).

The data reviewed earlier seemed inconclusive to permit one decide whether the protein synthesis theory of memory consolidation is a reasonable theory; many experimental results back it up while others challenge it. Three recent lines of evidence, however, seem to support the latter.

First, Canal et al. (2007) described a series of experiments in which anisomycin (ANI), a widely used PSI in memory consolidation studies, was infused into the amygdala after inhibitory avoidance training. A deficit of retention was found, as would be expected. More interesting, however, was the finding that release of norepinephrine, serotonin, and dopamine, in the amygdala, was higher than 1200%, 4000%, and 5000% relative to their basal levels, respectively, as revealed through microdialysis. Two years later, Qi and Gold (2009) reported similar effects after infusion of ANI into the hippocampus: a huge release of norepinephrine, dopamine, and acetylcholine. In both instances, appropriate additional experiments led to conclude that the amnesic effect of ANI was not due to the inhibition of protein synthesis but, rather, to the increased release of the neurotransmitters.

The second line of evidence that goes counter to the theory that protein synthesis is essential for memory consolidation stems from electrophysiological studies conducted by Sharma et al. (2012). They observed a drastic reduction of spontaneous slow-wave activity, of multiunit activity, and of the power spectra in the hippocampus after the administration of ANI.

The evidence provided by the these two analytic methodologies (microdialysis and electrophysiology) indicates that ANI (and very probably other PSIs) besides inhibiting protein synthesis produces drastic abnormalities that render the machinery of memory production incapable of performing any integrative activity that is thought to be necessary for memory formation. Common sense dictates that a learning experience should trigger a harmonious activation and inhibition of neuronal activity, inducing numerous functional interactions among sets of neurons and structures that eventually will lead to the development of a permanent memory store. This process can only be achieved with normal physiological electrical activity in neurons, which, among other things, allows for the timely release of minute quantities of neurotransmitters, each of which should appear within a particular time frame. If this exquisite synchrony of events is grossly disrupted, long-term memory is not feasible.

The third line of evidence against the protein theory of memory consolidation has to do with experiments that have used drugs that block the transcriptional and translational processes that are needed for new protein synthesis. There are numerous reports showing that infusions of ANI into specific cerebral structures induce amnesia (Quevedo et al., 1999); ANI halts translation, thus impeding the manufacture of new proteins. There is evidence that administration of drugs that arrest transcription, such as 5,6-dichloro-1-β-D-ribofuranosilbenzimidazol (DRB) thus impeding mRNA formation, also produce amnesia (Igaz et al., 2002).

A growing body of evidence shows that treatments that commonly produce amnesia become ineffective when administered to subjects that have been subjected to enhanced training (Prado-Alcala et al., 2012). Taking this evidence into account, we recently investigated if the protective effect of enhanced training against amnesic agents could also be found after blocking transcription and translation. To this end, ANI was bilaterally infused into the hippocampus of rats that had been trained with a foot-shock of either a relatively low or a relatively high intensity (enhanced training). The results showed that ANI produced an amnestic state in the low-foot-shock group and no memory deficiency at all in the enhanced-training group (Rodríguez-Serrano et al., 2009). When DRB was infused into the hippocampus of rats also trained with relatively low or high foot-shocks, amnesia and sparing of memory, respectively, was also seen (Torres-García et al., 2012). These two experiments strongly suggest that neither transcription (needed for new mRNA formation) nor translation (and, therefore, new protein synthesis) are needed for memory consolidation. This interpretation should be tentative because confirming data ought to be produced to test for the reliability of these results.

In sum, we have provided arguments, based on experimental data, that cast serious doubts about the validity and universality of the theory that memory consolidation is mediated by protein synthesis. This theory was first originated under the assumption that PSIs produce memory deficiencies precisely because of the inhibition of this synthesis. However, evidence has accumulated showing that PSIs have widespread unspecific effects that can reasonably account for the amnestic effects of PSIs.

Acknowledgments

Some experiments described in this article were financed by the National Council of Science and Technology of México (Grant CONACYT 128259) and by Dirección General de Asuntos del Personal Académico, Universidad Nacional Autónoma de México (Grant PAPIIT IN201712).

References

Agin, V., Chichery, R., Maubert, E., Chichery, M.P., 2003. Time-dependent effects of cycloheximide on long-term memory in the cuttlefish. Pharmacol. Biochem. Behav. 75, 141−146.

Agranoff, B.W., Klinger, P.D., 1964. Puromycin effect on memory fixation in the goldfish. Science 146, 952−953.

Agranoff, B.W., Davis, R.E., Brink, J.J., 1965. Memory fixation in the goldfish. Proc. Natl. Acad. Sci. USA. 54, 788−793.

Alberini, C.M., 2009. Transcription factors in long-term memory and synaptic plasticity. Physiol. Rev. 89, 121−145.

Atkins, C.M., Selcher, J.C., Petraitis, J.J., Trzaskos, J.M., Sweatt, J.D., 1998. The MAPK cascade is required for mammalian associative learning. Nat. Neurosci. 1, 602−609.

Bailey, C.H., Bartsch, D., Kandel, E.R., 1996. Toward a molecular definition of long-term memory storage. Proc. Natl. Acad. Sci. USA. 93, 13445−13452.

Barondes, S.H., Cohen, H.D., 1966. Puromycin effect on successive phases of memory storage. Science 151, 594−595.

Barondes, S.H., Cohen, H.D., 1967a. Comparative effects of cycloheximide and puromycin on cerebral protein synthesis and consolidation of memory in mice. Brain Res. 4, 44−51.

Barondes, S.H., Cohen, H.D., 1967b. Delayed and sustained effect of acetoxycycloheximide on memory in mice. Proc. Natl. Acad. Sci. USA. 58, 157−164.

Barraco, D.A., Lovell, K.L., Eisenstein, E.M., 1981. Effects of cycloheximide and puromycin on learning and retention in the cockroach, P. americana. Pharmacol. Biochem. Behav. 15, 489−494.

Bevilaqua, L.R.M., Kerr, D.S., Medina, J.H., Izquierdo, I., Cammarota, M., 2003. Inhibition of hippocampal Jun N-terminal kinase enhances short-term memory but blocks long-term memory formation and retrieval of an inhibitory avoidance task. Eur. J. Neurosci. 17, 897–902.

Blum, S., Moore, A.N., Adams, F., Dash, P.K., 1999. A mitogen-activated protein kinase cascade in the CA1/CA2 subfield of the dorsal hippocampus is essential for long-term spatial memory. J. Neurosci. 19, 3535–3544.

Bozon, B., Kelly, A., Josselyn, S.A., Silva, A.J., Davis, S., Laroche, S., 2003. MAPK, CREB and zif268 are all required for the consolidation of recognition memory. Philos. Trans. R. Soc. Lond. 358, 805–814.

Bull, R., Ferrera, E., Orrego, F., 1976. Effects of anisomycin on brain protein synthesis and passive avoidance learning in newborn chicks. J. Neurobiol. 7, 37–49.

Cammarota, M., Bevilaqua, L.R., Ardenghi, P.G., Paratcha, G., de Stein, M.L., Izquierdo, I., et al., 2000. Learning-associated activation of nuclear MAPK, CREB and Elk-1, along with Fos production, in the rat hippocampus after a one-trial avoidance learning: abolition by NMDA receptor blockade. Brain Res. Mol. Brain Res. 76, 36–46.

Canal, C.E., Chang, Q., Gold, P.E., 2007. Amnesia produced by altered release of neurotransmitters after intraamygdala injections of a protein synthesis inhibitor. Proc. Natl. Acad. Sci. USA. 104, 12500–12505.

Chevalier, J.A., 1965. Permanence of amnesia after a single posttrial electroconvulsive seizure. J. Comp. Physiol. Psychol. 59, 125–127.

Crow, T., Forrester, J., 1990. Inhibition of protein synthesis blocks long-term enhancement of generator potentials produced by one-trial in vivo conditioning in Hermissenda. Proc. Natl. Acad. Sci. USA. 87, 4490–4494.

Davis, H.P., Squire, L.R., 1984. Protein synthesis and memory: a review. Psychol. Bull. 96, 518–559.

DeZazzo, J., Tully, T., 1995. Dissection of memory formation: from behavioral pharmacology to molecular genetics. Trends Neurosci. 18, 212–218.

Díaz-Trujillo, A., Contreras, J., Medina, A.C., Silveyra-Leon, G.A., Antaramian, A., Quirarte, G.L., et al., 2009. Enhanced inhibitory avoidance learning prevents the long-term memory-impairing effects of cycloheximide, a protein synthesis inhibitor. Neurobiol. Learn. Mem. 91, 310–314.

Ennis, H.L., Lubin, M., 1964. Cycloheximide: aspects of inhibition of protein synthesis in mammalian cells. Science 146, 1474–1476.

Flexner, J.B., Flexner, L.B., Stellar, E., De La Haba, G., Roberts, R.B., 1962. Inhibition of protein synthesis in brain and learning and memory following puromycin. J. Neurochem. 9, 595–605.

Flexner, J.B., Flexner, L.B., Stellar, E., 1963. Memory in mice as affected by intracerebral puromycin. Science 141, 57–59.

Flexner, L.B., Flexner, J.B., 1966. Effect of acetoxycycloheximide and of an acetoxycycloheximide-puromycin mixture on cerebral protein synthesis and memory in mice. Proc. Natl. Acad. Sci. USA. 55, 369–374.

Flexner, L.B., Goodman, R.H., 1975. Studies on memory: inhibitors of protein synthesis also inhibit catecholamine synthesis. Proc. Natl. Acad. Sci. USA. 72, 4660–4663.

Flexner, L.B., Serota, R.G., Goodman, R.H., 1973. Cycloheximide and acetoxycycloheximide: inhibition of tyrosine hydroxylase activity and amnestic effects. Proc. Natl. Acad. Sci. USA. 70, 354–356.

Flood, J.F., Bennett, E.L., Rosenzweig, M.R., Orme, A.E., 1972. Influence of training strength on amnesia induced by pretraining injections of cycloheximide. Physiol. Behav. 9, 589—600.

Flood, J.F., Bennett, E.L., Orme, A.E., Rosenzweig, M.R., 1975. Effects of protein synthesis inhibition on memory for active avoidance training. Physiol. Behav. 14, 177—184.

Freedman, L.S., Judge, M.E., Quartermain, D., 1982. Effects of cycloheximide, a protein synthesis inhibitor, on mouse brain catecholamine biochemistry. Pharmacol. Biochem. Behav. 17, 187—191.

Freeman, F.M., Rose, S.P., Scholey, A.B., 1995. Two time windows of anisomycin-induced amnesia for passive avoidance training in the day-old chick. Neurobiol. Learn. Mem. 63, 291—295.

Goelet, P., Castellucci, V.F., Schacher, S., Kandel, E.R., 1986. The long and the short of long-term memory—a molecular framework. Nature 322, 419—422.

Gold, P.E., 2008. Protein synthesis inhibition and memory: formation vs amnesia. Neurobiol. Learn. Mem. 89, 201—211.

Guzowski, J.F., 2002. Insights into immediate-early gene function in hippocampal memory consolidation using antisense oligonucleotide and fluorescent imaging approaches. Hippocampus 12, 86—104.

Haycock, J.W., van Buskirk, R., Ryan, J.R., McGaugh, J.L., 1977. Enhancement of retention with centrally administered catecholamines. Exp. Neurol. 54, 199—208.

Hebb, D.O., 1949. The Organization of Behavior: A Neuropsychological Theory. John Wiley & Sons Inc., New York, NY.

Hraschek, A., Endroczi, E., 1978. Effects of systemic and intracerebral administration of adrenergic receptor blocking drugs on the conditioned avoidance behavior and maze learning in rats. Psychoneuroendocrinology 3, 271—277.

Igaz, L.M., Vianna, M.R., Medina, J.H., Izquierdo, I., 2002. Two time periods of hippocampal mRNA synthesis are required for memory consolidation of fear-motivated learning. J. Neurosci. 22, 6781—6789.

Izquierdo, I., Bevilaqua, L.R., Rossato, J.I., da Silva, W.C., Bonini, J., Medina, J.H., et al., 2008. The molecular cascades of long-term potentiation underlie memory consolidation of one-trial avoidance in the CA1 region of the dorsal hippocampus, but not in the basolateral amygdala or the neocortex. Neurotox. Res. 14, 273—294.

Jones, M.W., Errington, M.L., French, P.J., Fine, A., Bliss, T.V., Garel, S., et al., 2001. A requirement for the immediate early gene Zif268 in the expression of late LTP and long-term memories. Nat. Neurosci. 4, 289—296.

Katz, J.J., Halstead, W.C., 1950. Protein organization and mental function. Comp. Psychol. Monogr. 20, 1–38.

Lechner, H.A., Squire, L.R., Byrne, J.H., 1999. 100 Years of consolidation—remembering Muller and Pilzecker. Learn. Mem. 6, 77—87.

Luttges, M.W., McGaugh, J.L., 1967. Permanence of retrograde amnesia produced by electroconvulsive shock. Science 156, 408—410.

Martinez, J., Joel, L., Jensen, R.A., McGaugh, J.L., 1981. Attenuation of experimentally-induced amnesia. Prog. Neurobiol. 16, 155—186.

Mazzucchelli, C., Vantaggiato, C., Ciamei, A., Fasano, S., Pakhotin, P., Krezel, W., et al., 2002. Knockout of ERK1 MAP kinase enhances synaptic plasticity in the striatum and facilitates striatal-mediated learning and memory. Neuron 34, 807—820.

McGaugh, J.L., 1966. Time-dependent processes in memory storage. Science. 153, 1351—1358.

McGaugh, J.L., 2000. Memory—a century of consolidation. Science 287, 248—251.

Müller, G.E., Pilzecker, A., 1900. Experimentelle beiträge zur lehre vom gedächtnis. Zeitschrift für Psychologie 1, 1–288.

Prado-Alcala, R.A., Medina, A.C., Serafín, N., Quirarte, G.L., 2012. Intense emotional experiences and enhanced training prevent memory loss induced by post-training amnesic treatments administered to the striatum, amygdala, hippocampus or substantia nigra. Rev. Neurosci. 23, 501–508.

Qi, Z., Gold, P.E., 2009. Intrahippocampal infusions of anisomycin produce amnesia: contribution of increased release of norepinephrine, dopamine, and acetylcholine. Learn. Mem. 16, 308–314.

Quartermain, D., McEwen, B.S., Azmitia Jr., E.C., 1970. Amnesia produced by electroconvulsive shock or cycloheximide: conditions for recovery. Science 169, 683–686.

Quevedo, J., Vianna, M.R.M., Roesler, R., de Paris, F., Izquierdo, I., Rose, S.P.R., 1999. Two time windows of anisomycin-induced amnesia for inhibitory avoidance training in rats: protection from amnesia by pretraining but not pre-exposure to the task apparatus. Learn. Mem. 6, 600–607.

Ramon y Cajal, S., 1899. Textura del sistema nervioso del hombre y de los vertebrados. Imprenta y Librería de Nicolás Moya, Madrid.

Rodríguez-Serrano, L.M., Medina, A.C., Quirarte, G.L., Prado-Alcalá, R.A., 2009. Protein synthesis inhibition in dorsal hippocampus does not interfere with memory consolidation of enhanced inhibitory avoidance learning. Soc. Neurosci. Abstr. Chicago, IL, pp. 881.9.

Routtenberg, A., Rekart, J.L., 2005. Post-translational protein modification as the substrate for long-lasting memory. Trends Neurosci. 28, 12–19.

Sharma, A.V., Nargang, F.E., Dickson, C.T., 2012. Neurosilence: profound suppression of neural activity following intracerebral administration of the protein synthesis inhibitor anisomycin. J. Neurosci. 32, 2377–2387.

Sweatt, J.D., 2001. The neuronal MAP kinase cascade: a biochemical signal integration system subserving synaptic plasticity and memory. J. Neurochem. 76, 1–10.

Törocsik, B., Szeberenyi, J., 2000. Anisomycin uses multiple mechanisms to stimulate mitogen-activated protein kinases and gene expression and to inhibit neuronal differentiation in PC12 phaeochromocytoma cells. Eur. J. Neurosci. 12, 527–532.

Torres-García, M.E., Medina, A.C., Quirarte, G.L., Prado-Alcalá, R.A., 2012. Effects of post-trial injections of an inhibitor of RNAm synthesis in the dorsal hippocampus on memory consolidation of enhanced inhibitory avoidance. Soc. Neurosci. Abstr. New Orleans, LA, pp. 394.321.

Watanabe, H., Takaya, T., Shimoi, T., Ogawa, H., Kitamura, Y., Oka, K., 2005. Influence of mRNA and protein synthesis inhibitors on the long-term memory acquisition of classically conditioned earthworms. Neurobiol. Learn. Mem. 83, 151–157.

Yarmolinsky, M.B., Haba, G.L., 1959. Inhibition by puromycin of amino acid incorporation into protein. Proc. Natl. Acad. Sci. USA. 45, 1721–1729.

8 Basic Elements of Signal Transduction Pathways Involved in Chemical Neurotransmission

Claudia González-Espinosa and Fabiola Guzmán-Mejía

Pharmacobiology Department, Center for Research and Advanced Studies (Cinvestav), South Campus, Mexico City, México

Introduction

Research on cellular and molecular mechanisms involved in learning and memory constitutes a leading area in neuroscience that started long time ago. Since the mid-twentieth century, researchers have been concerned not only in finding the neurotransmitter systems involved in the transmission of electrical stimuli between nerve cells, but also in determining the long-term molecular changes that lead to long-term modifications to neural activity due to the neurotransmission process. One of the underlying hypothesis of molecular approaches to neurotransmission is that specific characteristics on the neuron-to-neuron communication (such as the intensity, duration, and composition of the signal) induce particular and sometimes irreversible modifications on neuron physiology leading to learning and memory consolidation. Description of the main neurotransmitter systems came together with the discovery of important changes in the postsynaptic terminals after the learning process. Those findings contributed to the late recognition of the importance of signal transduction events on the generation and maintenance of long-term changes on neural function, and allowed researchers to propose that abnormalities on particular signaling cascades could lead to modifications on the learning and memory processes.

Signal transduction systems modify the main intracellular information fluxes (Figure 8.1). Posttranslational protein modifications (such as phosphorylation, ubiquitination, and other), initiated by neurotransmitter−receptor interactions, alter intracellular protein activity, induce changes on gene transcription, and modify chromatin structure. Current research suggests those controlled modifications to neuron activity underlie learning and memory.

Multiple neurotransmitters have been linked to learning disorders and memory. Serotonin (Meneses and Ly-Salmerón, 2012), glutamate (Collingridge et al., 2013), and acetylcholine (Kosower, 1972) are the main examples of molecules that act on metabotropic and ionotropic receptors and start cascades able to produce soluble

Identification of Neural Markers Accompanying Memory. DOI: http://dx.doi.org/10.1016/B978-0-12-408139-0.00008-0

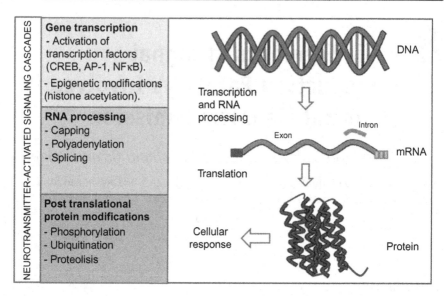

Figure 8.1 Signal transduction cascades modify the information fluxes inside the cell. Neurotransmitter–receptor interactions modify the activity of proteins and lead to chromatin changes that alter gene expression profiles. mRNA transcription and processing are intracellular events importantly regulated by memory and learning processes.

second messengers or modify, by phosphorylation and other chemical modifications, the activity of distinct proteins. Neurotransmitter receptors also induce changes on the activity of transcription factors and histone deacetylases, modifying gene expression and protein synthesis of particular groups of neurons.

The aim of this chapter is to introduce the reader to the general concepts of signal transduction, in order to provide some essential tools to understand the molecular mechanisms underlying the processes of learning and memory described in later chapters of this book.

Some Central Concepts on Cell-to-Cell Communication

Adaptability and survival of all organisms depends on the coordinated communication between cells and the external environment. Synaptic communication is defined as the interchange of chemical substances between neurons and, as in any other way of communication, requires three basic components: the signal transmitter, the signal transducer, and the signal receiver. This notion was initially proposed at the end of nineteenth century, when Santiago Ramon y Cajal conceived that neurons possessed three main regions: (i) the signal receiving region formed by the dendrites and soma, (ii) the transducer region, composed of the axon, and (iii) the signal emitting region

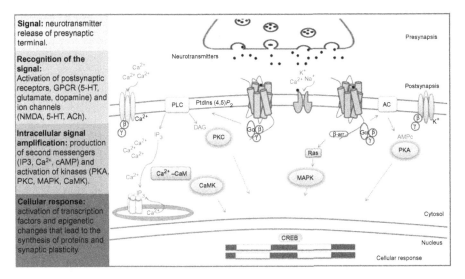

Figure 8.2 Main steps on signaling pathways involved in chemical neurotransmission. Neurotransmitters released in the presynaptic terminal activate specific receptors located in the postsynaptic neuron. Protein modification and second messengers are produced after receptor triggering, inducing the activation of transcription factors and chromatin modifications.

formed by terminal axonal or synaptic button (DeFelipe, 2010). To date, in signal transduction research in nervous system it is accepted that the signal emitting region is the presynaptic axon, the signal receiver region is any point of the postsynaptic neuron where neurotransmitter is released, and the transducer mechanism is composed by the receptors, ion channels, enzymes, transcription factors, and any biochemical element modified in the postsynaptic neuron, which lead to a correct decodification of the message and long-term changes on receptive neurons.

Some of the main steps in the process of chemical synaptic communication are depicted on Figure 8.2. Presynaptic terminal produces specific messengers that are released to the synaptic space and are recognized by particular receptors on the postsynaptic cell. This event leads to the activation of particular signaling pathways that involve the production of second messengers, activation of initial and intermediate kinases, nuclear translocation of transcription factors, and also to epigenetic modifications and changes on the translational machinery, which, in turn, will convert cell stimulation to long-term changes on gene expression profiles potentially related to learning and memory.

Neurotransmitters

Neurotransmitters and neuromodulators are the molecules responsible for the transmission of information on chemical synapses. For a molecule to be considered as a

Table 8.1 Main Neurotransmitters Associated with Learning and Memory, its Receptors and Canonical Signaling Pathways

Neurotransmitter	Receptors	Receptor Subtypes	Coupling
Acetylcholine	GCPR	M1 y M3	Gq
	Ion Channels	M2 y M4	Gi/o
		M5	Gq
		nAChR	Na^+
Adrenaline/noradrenaline	GCPR	B	Gs
		α1	Gq
		α2	Gi/o
Dopamine	GCPR	D1 (D_1y D_5)	Gs
		D2(D2S, D2L, D3, D4)	Gi/o
Serotonin	GCPR	$5HT_1$ y $5HT_5$	Gi/o
	Ion channels	$5HT_2$	Gq
		$5HT_4$, $5HT_6$ y $5HT_7$	Gs
		$5HT_3$	Na^+ y K^+
Histamine	GCPR	H1	Gq
		H2	Gs
		H3 y H4	Gi/o
Glutamate	GCPR	mGluR1 y 5	Gq
	Ion channels	mGluR2, 3,4,6,7,8	Gi/o
		AMPA	Na^+, K^+, Ca^{2+}
		Kainate	
		NMDA	
GABA	GCPR	GABA B	Gi/o
	Ion channels	GABA A and C	Cl^-

neurotransmitter (i) must be stored in vesicles together with the enzymes responsible for its synthesis; (ii) must be released in response to an increase in intracellular Ca^{2+}; and (iii) the exogenous administration of the neurotransmitter should elicit the same response as it were endogenously produced.

Neurotransmitters can be classified into two groups: (i) classic, such as amino acid derivatives and (ii) neuropeptides. The main neurotransmitters associated with learning and memory, together with its receptors and signaling systems, are given in Table 8.1.

G-Protein-Coupled Receptors

The G-protein-coupled receptors (GPCRs) represent a large and diverse family of molecules whose main function is to transmit information from the extracellular environment to the cell interior. These receptors are widely distributed throughout the central nervous system and are activated by various endogenous messengers, among which are, biogenic amines, hormones, peptides, proteins, growth factors, lipid, and ions (Lefkowitz, 2004). These receptors have been extensively studied in recent years, and they have been found more than 800 members in the human genome (Fredriksson et al., 2003), which make those receptors important targets for the therapy of many diseases.

GPCRs are high molecular weight proteins with seven transmembrane alpha helices. The seven transmembrane (TM 1–7) segments are connected by three extracellular and three intracellular loops (ECL1–ECL3 and ICL1–ICL3) (Venkatakrishnan et al., 2013). The amino-terminal region of the receptor is located on the extracellular, and the carboxy-terminal region is located in the cytoplasm.

GPCR Activation

Upon ligand binding, GPCRs suffer a conformational change able to activate different intracellular pathways dependent on G-proteins, β-arrestin, and other depending on GIPs (GPCR-interacting proteins) (Zheng et al., 2010).

The ICL and cytoplasmic regions of GPCRs are those involved in the coupling with different proteins. ILC1 generally consists of six amino acids that form a helical turn, while ICL2 forms a less structured domain with two helical turns and the binding region of protein G. ICl3 and the C-terminus are very large and variable regions that can bind to the β-arrestin, and the intracellular region is a short amphipathic helix (H8) which has been shown to facilitate the binding of protein G with ICL2 (Congreve et al., 2011).

G-Proteins

G-proteins are heterotrimeric proteins composed of three subunits (α, β, and γ). In humans, there have been identified several genes encoding isoforms of each subunit. According to the similarity between the structure and functionality of α subunits of G-proteins can be divided into four subfamilies, $G\alpha s$, $G\alpha q$, $G\alpha i/o$, and $G\alpha_{12/13}$, each coupled to a distinct signal transduction system.

$G\alpha$ subunit of G-proteins is normally bound to Guanosine-5-diphosphate (GDP) in the resting state, after receptor activation α subunits exchange the molecule GDP by Guanosine 5-triphosphate (GTP), leading to the dissociation of $G\alpha$ from $G\beta\gamma$ subunits (Woolf and Linderman, 2003).

Once dissociated, subunits $G\alpha$-GTP and $G\beta\gamma$ regulate the activity of effector enzymes, such as adenylate cyclase, phospholipase C (PLC), ion channels, and others, and promote the production of small molecules called "second messengers" (e.g., Ca^{2+}, cyclic $3'5'$-adenosine monophosphate (cAMP), inositol trisphosphate (IP3), and diacylglycerol (DAG)) (Rajagopal et al., 2010).

Receptor Desensitization

One of the most important characteristics of GPCR signaling is the process of turning off the signal, inducing receptor desensitization. After receptor activation, diverse kinases can phosphorylate and uncouple receptor form the downstream signaling system. It has been reported that protein kinase A (PKA), protein kinase C (PKC), and JAK3 are able to phosphorylate and silence signaling from GPCRs (Woolf and Linderman, 2003). A specific group of kinases, known as G protein-coupled receptor kinases (GRKs), phosphorylate threonine and serine residues on the third intracellular loop and the carboxyl terminus of the receptor. GRKs belong to a family of seven known members. GRK1 and 7 are expressed mainly in the retina, GRK4 is expressed in testis, and GRK2, 3, 5, and 6 are expressed in all tissues (Luttrell and Gesty-Palmer, 2010). GPCR sites phosphorylated by GRKs are recognized by small adapter molecules known as β-arrestins (Whalen et al., 2011).

The β-arrestins constitute a family of cytosolic protein scaffolds, and so far several members of the family have been identified. β-arrestin 1 (β-ARR1) is found exclusively on the retina while β-ARR2 and 3 are expressed ubiquitously. The β-arrestins possess a C-terminus domain able to interact with distinct proteins, such as clathrin, adaptin, and AP-2 complex, all of them are important participants in the endocytic machinery of the cell. That β-arrestin-containing complex favors the formation of an endocytic vesicle that is internalized in a dynamin and GTP-dependent fashion. Endocytosis removes β-arrestin-bound GPCRs from plasma membrane, but β-arrestin/GPCR complexes are still able to signal inside the cell. Internalized GPCR are recycled to the plasma membrane or can be degraded, depending on the β-arrestin and accessory molecules coupled to receptors (Maurice et al., 2011).

β-Arrestin-Dependent Signaling

For long time, it was considered that the β-arrestins functioned only as scaffold proteins that favored the process of desensitization of GPCRs. However, it is now well known that these proteins allow receptor to signal through different pathways, including those leading to the activation of the mitogen-activated kinases (MAPK), Src family kinases, the nuclear factor-κB (NFκB), and the phosphoinositide 3-kinase (PI3K).

Recent evidence indicates that β-arrestin regulates at least two members of the MAPK family, the regulatory kinase extracellular signal (ERK) kinase and the c-Jun N-terminal kinase (JNK). Also, β-ARR1 binds a number of plasma membrane proteins, including Src family kinases (such as Src and Fyn), leading to the phosphorylation of MAPKKK, MAPKK, MEK, and ERK (McDonald and Lefkowitz, 2001). Moreover, it has been shown that β-ARR2 can directly interact with MKK4 (Ask1), forming a complex that dramatically enhances JNK phosphorylation (Zheng et al., 2010).

Besides regulating MAPK after GPCR activation, β-arrestins regulate the activation of other receptors. For example, in cells lacking both isoforms of β-arrestin, the insulin-like factor receptor 1 (IGFR1) cannot stimulate PI3K. β-arrestins seem to be essential for PI3K activation, because when its expression is decreased, less Src and AKT activation is observed *in vivo* (Lefkowitz, 2004).

Signaling pathways controlled by β-arrestins are able to alter gene expression due to the activation of specific transcription factors, such as NFκB and cyclic

AMP response element-binding protein (CREB). Activation of those molecules can result of induced pathways or can be achieved by directed interaction between β-arrestins and NFκB or CREB. For example, both isoforms of β-arrestin can form a complex with the NFκB inhibitor (IκB) and this interaction stabilizes the NFκB−IκB complex resulting in inhibition of NFκB gene transcription. It has also been reported that β-ARR2 binds to promoter regions of specific genes, including p27 and Fos. βARR2 promotes the recruitment of histone acetyltransferase that interact with the p300 complex, favoring transcription (Rajagopal et al., 2010).

Small GTPases

Monomeric GTPases constitute a family of proteins able to bind GTP and hydro-lyze it changing from an inactive (GDP-bound) to an active (GTP-bound) state. These proteins have been grouped into five major families: Ras, Rho, Rab, Arf, and Ran, all regulated by various proteins such as guanine-nucleotide-exchange factors and GTPase-activating proteins that are activated after GPCR triggering (Etienne-Manneville and Hall, 2002). The main function of small GTPases is to couple receptor activation to changes on cytoskeleton and gene transcription (see MAPK section).

The Second Messengers

Second messengers refer to small intracellular molecules that are produced after the first messenger (hormone or neurotransmitter)-dependent receptor activation. Second messengers are intended to activate intracellular signaling pathways that amplify the signal and culminate with the activation or inhibition of transcription factors, inducing a cellular response.

The chemical nature of the second messenger is diverse: cyclic nucleotides, lipid derivatives and small active compounds, and some ions. The most studied second messengers are cyclic $3'5'$-adenosine monophosphate (AMP) or cyclic guanosine monophosphate (GMP), calcium, DAG, IP3, and reactive oxygen and nitrogen species (ROS, NOS). In the following sections, we emphasize on the general charac-teristics and mechanisms of action of some of them.

Cyclic Adenosine Monophosphate

cAMP is a low-molecular-weight, hydrophilic second messenger formed by adeno-sine trisphosphate hydrolysis by the action of the enzyme adenylyl cyclase (AC) located at the plasma membrane and is activated by G-proteins.

The class III AC is the most important member of the family. The structure of AC can be divided into five structural domains: the N-terminal cytoplasmic domain, the M1 and M2 domains formed by six transmembrane helixes each, and the C1 and C2 domains, which constitute the cytoplasmic catalytic site of the protein (Cooper, 2003). So far there have been described at least nine isoforms (type I to type IX) and are differentially regulated by both the α and $\beta\gamma$ subunits of G-proteins, which can activate or inhibit its activity (Ishikawa and Homcy, 1997).

Upon stimulation of AC, increased intracellular cAMP leads to the activation of a number of proteins, for example, cAMP can modulate the opening of some cationic channels that possess a binding domain for cyclic nucleotides (CNBD) (Biel, 2009). cAMP also activates the PKA.

PKA is conformed by two catalytic and two regulatory subunits. Each regulatory domain possesses a high affinity site for cAMP. Binding of cAMP to regulatory subunits induces a conformational change that allows this module to release the active catalytic subunits (Skroblin et al., 2010).

Catalytic subunits have a highly conserved core kinase domain in the N-terminal region, which binds ATP. The carboxy terminus of catalytic subunits binds a number of different substrates that phosphorylated at serine−threonine residues (Skroblin et al., 2010).

Inositol 1,4,5 Triphosphate

In addition to cyclic nucleotides, lipid-derived messengers also play an important role in signal transduction and amplification. The phosphatidylinositol 4,5 bisphosphate is a membrane lipid which is hydrolyzed by the enzyme PLC producing two important intracellular messengers, IP3 and DAG (Berridge and Irvine, 1984). PLC is a family of 13 different isoforms that are differentially regulated by multiple proteins, including Gαq (Smrcka et al., 1991) and G$\beta\gamma$ subunits (Camps et al., 1992). Some PLC isoforms possess structural domains that allow the recruitment of PLC to tyrosine-phosphorylated receptors and adapters (Smrcka et al., 2012).

IP3 is a water-soluble molecule which, when released from the plasma membrane by the action of PLC, diffuses into the cytoplasm and acts on specific tetrameric receptors located on the endoplasmic reticulum (ER). This interaction promotes the release of another important second messenger (Ca^{2+} ion) from intracellular stores (Parys and De Smedt, 2012).

DAG is produced by the action of PLC and constitutes an important activator of certain members of the PKC family of proteins, which posses binding domains for DAG and Ca^{2+} ions and promotes the phosphorylation of other proteins and amplifying the signal (Newton, 2010).

Calcium

Calcium as a second messenger that possesses certain peculiarities, first Ca^{2+} localized at the cytoplasm may come from different sources, either intracellular stores or the extracellular space. Calcium from intracellular stores is released by the activation of IP3 or ryanodine receptors located at intracellular membranes (such as ER), whereas extracellular Ca^{2+} enters to cells through specific channels divided in three main categories: (i) voltage-operated channels (VOCs), (ii) receptor-operated channels (ROCs), and (iii) store operated channels (SOCs) (Berridge et al., 1998).

Cells are able to detect very slight changes in the concentrations of cytoplasmic Ca^{2+} and couple those changes to particular activation of regulatory proteins and transcription factors, regulating cell motility, apoptosis, protein synthesis, and exocytosis, among other functions (Berridge et al., 2000).

Distinct proteins in cell cytosol are able to bind Ca^{2+} modifying its activity. For example, calmodulin and calcineurin possess EF-hand domains that bind Ca^{2+}, while PKC and the vesicle protein synaptotagmin present C2–Ca^{2+} binding domains. On its side, PLC contains both (EF-hand and C) domains to bind Ca^{2+}.

Calmodulin (CaM) contains two EF–Ca^{2+} domains connected by an alpha-helix loop. Upon Ca^{2+} binding, CaM suffers a conformational change and binds to multiple proteins (Saucerman and Bers, 2012) that are then activated. For example, Ca^{2+}–CaM is able to bind to the calcium–calmodulin kinase IV (CaMKIV), which is forming an inactive complex in cell cytoplasm. After Ca^{2+}–CaM binding to CaMKIV, the latter dissociates from its inhibitor and becomes able to phosphorylate distinct intracellular proteins (Means, 2008).

Other molecule able to bind Ca^{2+} is the phosphatase calcineurin, that is a serine/threonine phosphatase composed by one catalytic (A) and one regulatory (B) subunit. Occupation of the calcium-binding site at the B subunit induces a conformational change that exposes a Ca^{2+}–CaM binding site on A subunit. Once Ca^{2+}–CaM binds to A subunit, it becomes active, dephoshphorylating a number of intracellular targets (Li et al., 2011).

Synaptotagmin is a protein involved in exocytosis of neurotransmitter-containing vesicles present in all neurons. This protein contains two Ca^{2+} binding (C2) domains. Calcium bound to C2 domains does not induce conformational changes but alters electrostatic properties of proteins. Coalition of Ca^{2+}-containing synaptotagmin increases its affinity for hydrophobic peptides and lipids, being now able to binds AP-2 and SNARE proteins, favoring the secretory process (Sudhof, 2002).

Mitogen-Activated Protein Kinases

After the production of second messengers and activation of initial kinases (such as PKA, PKC, and CAMKIV), signaling pathways lead to the activation of another group of intermediate enzymes, able to integrate signals coming from different receptors. Members of this late group of enzymes are known as the mitogen-activated protein kinases (MAPKs).

MAPK activation requires phosphorylation of threonine and tyrosine residues of the catalytic core, but also can exert some functions through a domain named anchor region that interact with phosphatases or other substrates in cytoplasm and nucleus (Yang et al., 2013). So far, four major MAPKs have been described: the extracellular signal-regulated kinase 1/2 (ERK), JNK, p-38, and ERK5 (Plotnikov et al., 2011).

Activation of the MAPK cascade occurs in a sequential series of events that involve the participation of three elements: (i) MAP kinase kinase kinase (MAPKKK), (ii) MAP kinase kinase (MAPKK), and (iii) MAP kinase (MAPK) (Yang et al., 2013). The sequence of events from membrane receptor activation to specific MAPK triggering is a matter of intense research work, and it is known that activation of ERK1/2 occurs after a series of reactions in which monomeric GTPase (Ras) activates specific MAPKK (MEK1 and MEK2), which ultimately phosphorylate ERK1/2 serine residues. Phosphorylated ERK is capable of activating transcription factors and enzymes involved in posttranslational modification (PTM) of histones (Brunner et al., 2012). The cascade leading to p-38 and JNK activation has high similarity to ERK triggering, with to some variants on the MAPKKK and the MAPKK needed for their phosphorylation.

The Transcription Factors

Transcription factors are proteins that control the expression of multiple genes. These proteins are able to bind to specific sequences that are located in the promoter region of each gene, and this binding favors the coupling of other proteins that promote or block recruitment of RNA polymerase and transcription machinery.

CREB is one of the transcription factors best associated to learning and memory (Kim et al., 2013). It has been largely probed that CREB transcription factor is fundamental for long-term memory consolidation. For example, in a study performed with a murine model of passive avoidance, it was found that training induced high levels of CREB phosphorylation in CA1 and CA3 areas of hippocampus. In other models, such as the recognition of new objects, it has been observed the accumulation of high quantities of phosphorylated CREB also in hippocampus (Xia and Storm, 2012). To date, it is known that multiple signals can induce phosphorylation of CREB, including membrane depolarization, an increased intracellular cAMP, the MAPK kinases-dependent Ca^{2+}-CaM (West et al., 2002).

CREB family includes CREB, CREM (CRE-modulating protein), and ATF1 (activating transcription factor). All members of this family possess an activation N-terminal domain and a C-terminal DNA binding and dimerization domain. In the amino-terminal region, or transactivation domain, there are two specific motifs, one for constitutive activation (Q2) and one named kinase-inducible domain (KID) (Altarejos and Montminy, 2011). KID domain consists of about 58 amino acids and contains the site for PKA phosphorylation at serine 133.

After stimulation and increased intracellular cAMP, catalytic subunit of PKA translocates to cell nucleus in approximately 20 min and attenuation of the response occurs between 4 and 6 h after stimulation with the agonist. The attenuation process involves the participation of other proteins such as PP-1 (a serine/threonine phosphatase). PP-1 is bound to chromatin associated with its inhibitor (NIPP-1), that is a target for PKA. When the catalytic subunit of PKA phosphorylates NIPP-1, it dissociates from PP-1, increasing its phosphatase activity and stopping the activity of CREB. Finally, after a long period of stimulation downregulation of the catalytic subunit of PKA occurs together with an increase in phosphatase activity of PP-1 (Montminy, 1997), desensitizing the system.

PKA-dependent CREB phosphorylation at serine 133 promotes its association with CBP (CREB-binding protein). This complex leads to the improvement of transcription because it increases histone acetylation and promotes the recruitment of RNA polymerase. Other positive modulators are the CRTC (CREB-regulated transcription coactivators). In basal conditions, CREB/CRTC complexes are localized in the cytoplasm but after increasing intracellular Ca^{2+} or cAMP, calcineurin dephosphorylate them, allowing their nuclear translocation (Altarejos and Montminy, 2011).

Epigenetic Modifications

Gene transcription is controlled by complex interactions between transcription factors, histones, and proteins that modify chromatin structure. Neurotransmitter—receptor activation causes activation of specific enzymes that generate PTMs on histones, such as acetylation and phosphorylation (Brunner et al., 2012). Besides, signaling pathways also induce changes on methylation patterns of DNA, repressing gene expression (Gos, 2013).

Future research on epigenetic modifications induced by the process of neurotransmission will help to understand the molecular changes associated with learning and memory.

References

Altarejos, J.Y., Montminy, M., 2011. CREB and the CRTC co-activators: sensors for hormonal and metabolic signals. Nat. Rev. Mol. Cell Biol. 12 (3), 141−151.

Berridge, M.J., Irvine, R.F., 1984. Inositol trisphosphate, a novel second messenger in cellular signal transduction. Nature 312 (5992), 315−321.

Berridge, M.J., Bootman, M.D., Lipp, P., 1998. Calcium—a life and death signal. Nature 395 (6703), 645−648.

Berridge, M.J., Lipp, P., Bootman, M.D., 2000. The versatility and universality of calcium signalling. Nat. Rev. Mol. Cell Biol. 1 (1), 11−21.

Biel, M., 2009. Cyclic nucleotide-regulated cation channels. J. Biol. Chem. 284 (14), 9017−9021.

Brunner, A.M., Tweedie-Cullen, R.Y., Mansuy, I.M., 2012. Epigenetic modifications of the neuroproteome. Proteomics 12 (15−16), 2404−2420.

Camps, M., Carozzi, A., Schnabel, P., Scheer, A., Parker, P.J., Gierschik, P., 1992. Isozyme-selective stimulation of phospholipase C-beta 2 by G protein beta gamma-subunits. Nature 360 (6405), 684−686.

Collingridge, G.L., et al., 2013. The NMDA receptor as a target for cognitive enhancement. Neuropharmacology 64, 13–26.

Congreve, M., et al., 2011. Progress in structure based drug design for G protein-coupled receptors. J. Med. Chem. 54 (13), 4283–4311.

Cooper, D.M., 2003. Regulation and organization of adenylyl cyclases and cAMP. Biochem. J. 375 (Pt 3), 517–529.

DeFelipe, J., 2010. From the connectome to the synaptome: an epic love story. Science 330 (6008), 1198–1201.

Etienne-Manneville, S., Hall, A., 2002. Rho GTPases in cell biology. Nature 420 (6916), 629–635.

Fredriksson, R., et al., 2003. The G-protein-coupled receptors in the human genome form five main families. Phylogenetic analysis, paralogon groups, and fingerprints. Mol. Pharmacol. 63 (6), 1256–1272.

Gos, M., 2013. Epigenetic mechanisms of gene expression regulation in neurological diseases. Acta Neurobiol. Exp. (Wars). 73 (1), 19–37.

Ishikawa, Y., Homcy, C.J., 1997. The adenylyl cyclases as integrators of transmembrane signal transduction. Circ. Res. 80 (3), 297–304.

Kim, J., et al., 2013. CREB and neuronal selection for memory trace. Front. Neural Circuits 7, 44.

Kosower, E.M., 1972. A molecular basis for learning and memory. Proc. Natl. Acad. Sci. USA. 69 (11), 3292–3296.

Lefkowitz, R.J., 2004. Historical review: a brief history and personal retrospective of seven-transmembrane receptors. Trends Pharmacol. Sci. 25 (8), 413–422.

Li, H., Rao, A., Hogan, P.G., 2011. Interaction of calcineurin with substrates and targeting proteins. Trends Cell Biol. 21 (2), 91–103.

Luttrell, L.M., Gesty-Palmer, D., 2010. Beyond desensitization: physiological relevance of arrestin-dependent signaling. Pharmacol. Rev. 62 (2), 305–330.

Maurice, P., et al., 2011. GPCR-interacting proteins, major players of GPCR function. Adv. Pharmacol. 62, 349–380.

McDonald, P.H., Lefkowitz, R.J., 2001. Beta-Arrestins: new roles in regulating heptahelical receptors' functions. Cell Signal. 13 (10), 683–689.

Means, A.R., 2008. The year in basic science: calmodulin kinase cascades. Mol. Endocrinol. 22 (12), 2759–2765.

Meneses, A., Ly-Salmerón, G., 2012. Serotonin and emotion, learning and memory. Rev. Neurosci. 23, 543–553.

Montminy, M., 1997. Transcriptional regulation by cyclic AMP. Annu. Rev. Biochem. 66, 807–822.

Newton, A.C., 2010. Protein kinase C: poised to signal. Am. J. Physiol. Endocrinol. Metab. 298 (3), E395–E402.

Parys, J.B., De Smedt, H., 2012. Inositol 1,4,5-trisphosphate and its receptors. Adv. Exp. Med. Biol. 740, 255–279.

Plotnikov, A., et al., 2011. The MAPK cascades: signaling components, nuclear roles and mechanisms of nuclear translocation. Biochim. Biophys. Acta 1813 (9), 1619–1633.

Rajagopal, S., Rajagopal, K., Lefkowitz, R.J., 2010. Teaching old receptors new tricks: biasing seven-transmembrane receptors. Nat. Rev. Drug Discov. 9 (5), 373–386.

Saucerman, J.J., Bers, D.M., 2012. Calmodulin binding proteins provide domains of local Ca^{2+} signaling in cardiac myocytes. J. Mol. Cell Cardiol. 52 (2), 312–316.

Skroblin, P., et al., 2010. Mechanisms of protein kinase a anchoring. Int. Rev. Cell Mol. Biol. 283, 235–330.

Smrcka, A.V., et al., 1991. Regulation of polyphosphoinositide-specific phospholipase C activity by purified Gq. Science 251 (4995), 804−807.

Smrcka, A.V., Brown, J.H., Holz, G.G., 2012. Role of phospholipase Cepsilon in physiological phosphoinositide signaling networks. Cell Signal. 24 (6), 1333−1343.

Sudhof, T.C., 2002. Synaptotagmins: why so many? J. Biol. Chem. 277 (10), 7629−7632.

Venkatakrishnan, A.J., et al., 2013. Molecular signatures of G-protein-coupled receptors. Nature 494 (7436), 185−194.

West, A.E., Griffith, E.C., Greenberg, M.E., 2002. Regulation of transcription factors by neuronal activity. Nat. Rev. Neurosci. 3 (12), 921−931.

Whalen, E.J., Rajagopal, S., Lefkowitz, R.J., 2011. Therapeutic potential of beta-arrestin- and G protein-biased agonists. Trends Mol. Med. 17 (3), 126−139.

Woolf, P.J., Linderman, J.J., 2003. Untangling ligand induced activation and desensitization of G-protein-coupled receptors. Biophys. J. 84 (1), 3−13.

Xia, Z., Storm, D.R., 2012. Role of signal transduction crosstalk between adenylyl cyclase and MAP kinase in hippocampus-dependent memory. Learn. Mem. 19 (9), 369−374.

Yang, S.H., Sharrocks, A.D., Whitmarsh, A.J., 2013. MAP kinase signalling cascades and transcriptional regulation. Gene 513 (1), 1−13.

Zheng, H., Loh, H.H., Law, P.Y., 2010. Agonist-selective signaling of G protein-coupled receptor: mechanisms and implications. IUBMB Life 62 (2), 112−119.

9 A Role for Learning and Memory in the Expression of an Innate Behavior: The Case of Copulatory Behavior

Gabriela Rodríguez-Manzo and Ana Canseco-Alba

Pharmacobiology Department, Center for Research and Advanced Studies (Cinvestav), South Campus, Mexico City, Mexico

Introduction

Animal behavior can be broadly divided into instinctive or innate behaviors and learned behaviors. Instinctive behaviors are genetically determined; they are stereotyped (e.g., they are executed the same way every time); their expression requires no experience, meaning that they manifest without prior learning and consequently, memory is not important for its display. Learned behaviors, by contrast, are acquired behaviors; animals are not born with them; their performance is improved by experience and memory plays an important role in its expression. Learned behavior can be changed and supposedly instinctive behavior cannot. However, some innate behaviors may be modified (or modulated) through practice and experience, indicating that the classification of behaviors in either "innate" or "learned" might be excessive. Most behaviors are a mix of the two, neither completely innate nor entirely learned (for review see Barnard, 2004; Bolhuis and Giraldeau, 2005). This is the case of sexual behavior. In addition, instinctive and learned behaviors share a common neural substrate as evidenced by the fact that lesion of specific brain areas impairs both learned and instinctive behaviors. Thus, experimental destruction of the neocortex or the hippocampus in laboratory rats disrupts running through a maze without making errors and impairs the performance of sexual behavior (Vanderwolf, 2003).

This chapter is aimed to review evidence for a role of learning and memory in diverse aspects of a behavior that is classified as instinctive: copulatory behavior.

Learning can be defined as an adaptive change in behavior caused by experience and the storage and recall of previous experiences describes the term memory (Kandel and Squire, 2009). Animals must be able to make adjustments in their nervous systems in response to environmental stimuli that permit adaptive behavioral responses (Barnard, 2004). The ability of animals to adapt behaviorally in response

Identification of Neural Markers Accompanying Memory. DOI: http://dx.doi.org/10.1016/B978-0-12-408139-0.00009-2

to external stimuli is essential for survival. For a long time it was considered that this ability was not involved in the display of instinctive, hard-wired behaviors which only reflected the activation of developmentally programmed neural circuits that were not modifiable by the environment.

Sexual activity is an innate highly rewarding behavior. According to Thorndike's Law of Effect (1911) a reward increases the frequency and intensity of a specific behavioral act that has resulted in a reward before (Schultz, 2006). The ability to learn about rewards is also a crucial adaptive capacity and the ability to learn and to remember what is learned is important to maintain adaptations (Barnard, 2004; Bolhuis and Giraldeau, 2005). The rewarding component of sexual activity ensures its repetition contributing to guarantee species survival. The appropriate expression of an innate behavioral sequence frequently requires signals from the outside world (Burkhardt, 2005). A key component of reward-related learning is the acquisition of conditioned associations between rewards and environmental stimuli that predict or accompany those rewards. Such stimuli are called "conditioned stimuli" (Zellner and Ranaldi, 2010). The rewarding and reinforcing character of male rat sexual behavior is evidenced by the fact that male rats form a conditioned place preference (CPP) for copulation (Ågmo and Berenfeld, 1990; Pfaus and Phillips, 1991; Tenk et al., 2009), demonstrating that the male rat is able to associate an environment with the rewarding sexual experience and is capable of recalling that association.

Sexual behavior is an innate behavior which expression is delayed until puberty. During this period important hormonal variations underlie sexual maturation, which include hormonal, anatomical, and neurochemical changes in all sex-related tissues, including the brain (Schulz et al., 2009). The development of copulatory behavior in rats depends on gonadal hormones, since its appearance is prevented by prepubertal castration (Larsson, 1967). Puberty-related brain changes trigger sexual behavior display in response to the adequate stimuli, thus male sexual behavior expression is regulated by brain circuits that are importantly shaped by sex hormones such as testosterone and estrogen (Simerly, 2005). Although sexual development is determined by steroid hormonal exposure, learning from social and sexual experiences at different stages contributes to the shaping of sexual function (Woodson, 2002). Sex hormones participate also in neuronal plasticity processes in the adult brain (Dugger et al., 2008); therefore, such neural plasticity might be one of the mechanisms underlying the learning processes associated to sexual behavior expression in adult individuals.

Male Rat Sexual Behavior

The copulatory pattern of the male rat consists of a stereotyped behavioral sequence composed by three distinct motor behaviors: mounting, intromissive, and ejaculatory. During mounting behavior the male rat typically poses his forelegs over the female's back and the sexually receptive female may respond with the lordosis posture: a dorsiflexion of the spine that is accompanied by a flexion of the tail to the side. Then, with his hind feet on the ground, the male rat begins anteroposterior pelvic thrusting without achieving penile insertion into the female's vagina; once ended the male dismounts the female slowly. Intromissive behavior implies intravaginal penile insertion. During this motor response the male mounts the female, performs pelvic thrusting and displays a deeper thrust that coincides with vaginal penetration, followed by an abrupt backward dismount and grooming of his genitalia. Ejaculatory behavior starts with an intromission, during which a deep, long thrust coincides with seminal ejection. The male rat then raises his forelegs and dismounts the female slowly, sometimes falling on the side, and typically grooms himself. These motor responses are integrated in the so-called rat copulatory pattern consisting of a number of successive mounts, interleaved with intromissions until the male achieves the ejaculatory threshold and ejaculates, exhibiting the associated ejaculatory behavior. After ejaculation, the rat goes into a refractory period during which it does not respond to the presence of the receptive female rat. At the end of this refractory period the animal is able to reinitiate copulation and resumes pursuance and mounting of the female (for review see Hull and Rodríguez-Manzo, 2009). The stereotyped character of this copulatory pattern allowed the establishment of temporal and numerical measures used to describe and evaluate the quality of mating behavior (Larsson, 1956; Dewsbury, 1979). The copulatory patterns displayed by males of other rodents species like mice, hamsters, and guinea pigs are very similar to the one here described for rats, although there are important differences in the details of the different motor behaviors, for example, their duration and their number (Hull and Dominguez, 2007).

Effects of Sexual Experience on Copulatory Behavior Expression

Sexual experience causes facilitation of subsequent sexual behavior (Pfaus et al., 2001). Sexually naïve male rats notably require a long time to investigate the female partner, sniffing and licking the anogenital region before initiating mating. In addition, these inexperienced males display a high number of mounts and intromissions and need a long time to achieve ejaculation. Previous sexual experience improves the efficiency of copulation, since repeated exposure to receptive females gradually reduces latencies to mount, intromit, and ejaculate, and diminishes the number of mounts and intromissions that precede ejaculation (Larsson, 1956). This improvement results in a stable copulatory pattern, resistant to environmental changes (Pfaus and Wilkins, 1995). The rewarding nature of sexual experience contributes to reward-related learning that seems to play a role in the stabilization of sexual behavior expression.

All the facets of sexual activity are rewarding, yet the distinct components of copulatory behavior seem to have a different rewarding value, with ejaculation being the most reinforcing component (Pfaus and Phillips, 1991). For instance, ejaculation is essential for the formation of CPP; males executing only mounts and intromissions without ejaculation do not form a CPP (Kippin and Pfaus, 2001). Interestingly, sexual experience modifies the rewarding value of the different components of the copulatory pattern. Thus, it has been reported that intromissions are capable of inducing CPP in animals without sexual experience (i.e., sexually naïve) and not in animals that experienced copulation to ejaculation (Tenk et al., 2009). Moreover, repeated noncopulatory exposures to an estrous female can also enhance copulation on the first sexual experience in sexually naïve males (Lagoda et al., 2004; Powell et al., 2003; Vigdorchik et al., 2012). These data show that there is a hierarchy of rewarding sexual behavior, with ejaculation being the most rewarding component, and that the rewarding incentive value of other sexual behavior components is dependent upon prior sexual experience (Tenk et al., 2009). The strength of sex as a reinforcer is directly related to the extent to which subjects complete the copulatory behavioral sequence (Crawford et al., 1993). In line with this data, to be able to develop the preference for a female rat's estrous odor, male rats require sexual experience (Hosokawa and Chiba, 2005; Ballard and Wood, 2007).

Experience plays an important role in the full development and efficiency of sexual behavior. Some of the learning-induced changes in male sexual ability can be interpreted as experience-dependent increases in sexual interest, since males engage in sexual behavior more rapidly, ejaculate more often, and display shorter refractory periods after ejaculation (Dewsbury, 1979; Larsson, 1956). Sexual experience-related development of the preference for an estrous female odor may be a reflection of an anticipation of mating, conditioned by prior sexual experience (Hosokawa and Chiba, 2005). Also, the intensified investigation of female chemosignals exhibited by sexually experienced male mice has been suggested to reflect their increased sexual interest as compared to sexually naïve animals (Swaney et al., 2007). An exception to these results is seen in a few individuals of all strains of rats, in which repeated sexual experience does not reduce the long latencies to initiate copulation and to achieve ejaculation; these animals are called "sluggish" male rats (Figure 9.1). At present, there is no clear explanation for the "lack" of experience-driven improvement in copulation of sluggish male rats, but the possible involvement of a learning impairment should be considered.

Another effect of sexual experience on sexual behavior expression is that it diminishes or eliminates the disruptive effect of external factors and physiological insults that occur in males that have not experienced sexual activity (sexually naïve males). For instance, in sexually naïve rats the exposure to a novel environment is a stressful experience that disrupts copulatory behavior and results in a reduction in the percentage of animals that engage in copulation and the sexual performance of those copulating is deficient. Sexual experience appears to diminish or even eliminate these disruptive effects of novelty (Pfaus and Wilkins, 1995).

Also, sexually experienced male mice (Manning and Thompson, 1976) and hamsters (Constantini et al., 2007), but not rats (Bloch and Davidson, 1968), are less susceptible to the disruptive effects of castration and show sexual behavior for longer periods after surgery than sexually naïve animals. The negative effects

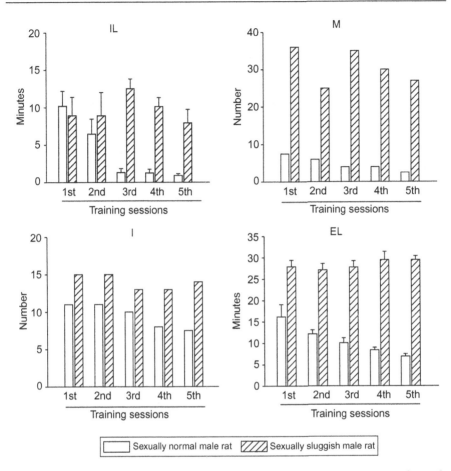

Sexually normal male rat ▢ Sexually sluggish male rat ▨

Figure 9.1 Improvement of specific sexual behavior parameters as a consequence of sexual experience. This figure illustrates the evolution of the different components of the copulatory pattern of sexually naïve male rats along successive sexual behavior training sessions with estrous females. Empty bars depict the behavior of the distinct specific sexual behavioral parameters in normal animals and dashed bars that of the so-called sexually sluggish males (see text). The recorded parameters were the time that the animals need to engage in sexual activity expressed through the intromission latency (IL), the amount of stimulation that males need to achieve the ejaculatory threshold expressed by the number of mounts (M) and intromissions (I) preceding ejaculation, and the time required by animals to achieve ejaculation expressed as the ejaculation latency (EL). Notice that normal animals (empty bars) show progressive reductions in all parameters until their stabilization which is achieved between the 3rd and 4th training sessions. Sexually sluggish males (dashed bars) by contrast do not exhibit reductions in any sexual behavior parameter, evidencing that sexual experience does not improve their sexual performance.

of specific brain lesions on sexual performance are also attenuated by sexual experience. Lesions affecting sexual stimuli perception such as bilateral olfactory bulbectomy in rats (Bermant and Taylor, 1969) and ablation of the vomeronasal organ (VNO) in hamsters (Meredith, 1986) and in rats (Saito and Moltz, 1986) depress sexual arousal, but this depression is more pronounced in sexually naïve than in sexually experienced males. Similarly, the effects of lesions in brain areas that are central for sexual behavior expression, like the medial preoptic area (mPOA) and the medial posterior bed nucleus of the stria terminalis (BNST) (Hull and Rodríguez-Manzo, 2009), were also greater in sexually naïve than in sexually experienced male rats (Arendash and Gorski, 1983; Claro et al., 1995; de Jonge et al., 1989).

Effects of Sexual Experience on Brain Functioning

There is evidence that sexually experienced animals exhibit several differences in brain functioning as compared to sexually naïve rats. For instance, it has been reported that sexual experience enhances neuronal responses. Thus, the odors from estrous females are detected by receptors in the VNO, which is the first segment of the vomeronasal pathway that ends both in the BNST and the mPOA. In male rats, Fos protein, the product of the immediate early gene c-fos used to identify neuronal activation, is induced after mating in all the segments of the vomeronasal pathway (Robertson et al., 1991). Increases in Fos expression in response to estrous odors are detected in the mPOA, the BNST, and the nucleus accumbens (NAcc), which is involved in sexual reward of sexually experienced (Mitchell and Gratton, 1994), but not of sexually naïve animals (Hosokawa and Chiba, 2005). In sexually experienced male rats, ejaculation activates more cells within the mPOA (Lumley and Hull, 1999) and, after exposure to an estrous female without physical contact, sexually experienced males show higher Fos expression in the NAcc than sexually naïve males (Lopez and Ettenberg, 2002).

Besides, chronic sexual experience promotes cell proliferation and neurogenesis, and increases the number of dendritic spines in the dentate gyrus of adult male rats (Leuner et al., 2010). Sexually experienced rats also exhibit reduced anxiety-like behaviors (Edinger and Frye, 2007) and increased sensitivity to drug actions (Rodríguez-Manzo et al., 2011) as compared to sexually naïve animals. All these data evidence the occurrence of brain plasticity changes as a result of sexual experience.

In line with this idea, Olsen states that natural reinforcers, like sexual reward, are capable of leading to plasticity in behavior and neurotransmission, in order to accomplish the required adaptation in behavior (Olsen, 2011).

Exploring the molecular mechanisms underlying the neuroadaptations that follow sexual experience, Bradley et al. (2005) found an increased expression of a large number of genes in the NAcc and dorsal striatum of sexually experienced female hamsters as compared to sexually naïve females, in response to a mating test.

Brain Regions Involved in Sexual Experience-Induced Behavioral Changes

The mesocorticolimbic dopaminergic system, originating in the ventral tegmental area (VTA) and projecting mainly to the NAcc, basolateral amygdala (BLA) and prefrontal cortex (PFC), plays an important role in mediating the behavioral effects of primary rewards (Berridge and Robinson, 1998). Early studies identified dopamine (DA) as an important neurotransmitter in operant responding to natural rewards (Wise et al., 1978) and sexual behavior elevates mesocorticolimbic DA (Damsma et al., 1992, Fiorino et al., 1997; Pfaus et al., 1990). Natural reward-related learning and memory are part of the behavioral adaptations induced by mating behavior experience (Pitchers et al., 2010b). It has been suggested that learned associations between sexual opportunities and conditions which maximize partner availability is not only beneficial to survival but also a natural function of the brain (Alcock, 2005).

DA at the mesocorticolimbic system has been proposed to play a role in synaptic plasticity and memory (Jay, 2003). The neural structure of the mesocorticolimbic system is altered by sexual activity (Fiorino and Kolb, 2003; Pitchers et al., 2010a). Thus, copulation increases dendritic arborization and the number of spines in the NAcc (Pitchers et al., 2010a); activates plasticity-related signaling cascades (Bradley et al., 2005; Hedges et al., 2009; Pitchers et al., 2010a) such as Mitogen-activated protein (MAP) kinases and Delta FosB, and causes long-term alterations in glutamate receptor expression and function in the NAcc (Pitchers et al., 2012). In addition, repeated sexual encounters sensitize the NAcc DA response to a later encounter (Kohlert and Meisel, 1999).

The behavioral changes induced by mating behavior could be mediated by alterations in the mesocorticolimbic system (Pitchers et al., 2010a). A mechanism that has been suggested to underlie reward-related learning involves the activation of VTA DA neurons. The suggested mechanism implies that the coincident stimulation from afferents carrying primary reward signals and from afferents carrying environmental stimuli signals would trigger neural plasticity in the VTA, mediated by N-methyl-D-aspartate (NMDA) receptors, which in turn would allow the activation of dopaminergic neurons (Zellner et al., 2009; Zellner and Ranaldi, 2010). The authors of this proposal believe that one of the consequences of the activation of the mesocorticolimbic system is to facilitate the associative processes underlying learning, allowing stimuli to guide future behavior.

VTA neurons are activated by a range of reward-related stimuli. Sexual reward in particular, has been found to increase Fos immunoreactivity in the VTA of rats when presented with a conditioned stimulus associated to mating (Coria-Avila and Pfaus, 2007). The VTA receives a confluence of acetylcholine and glutamatergic signals, which could represent some of the signals required for reward-related learning of both unconditioned and conditioned stimuli (Zellner and Ranaldi, 2010).

Recently, McHenry et al. (2012) proposed that DA receptors of the D1 family and the phosphorylation of the dopamine- and cyclic AMP-regulated phosphoprotein-32 (DARPP-32) in the mPOA, are involved in the enhancement of male rat sexual behavior induced by sexual experience. This proposal is based on the fact that DA acting at D1-like receptors facilitates male sexual behavior (Markowski et al., 1994), associative learning (El-Ghundi et al., 2007) and synaptic plasticity (Yao et al., 2008). In addition, D1 receptor actions in the NAcc appear to contribute to the experience-induced enhancement of male sexual behavior (Bialy et al., 2010). Thus, it seems that DA release at multiple brain areas in response to both copulation and/or sex-related stimuli act at D1-like receptors in each of these brain regions to collectively contribute to the enhancement of sexual behavior. NMDA glutamate receptors interact with D1 receptors in other brain regions to synergistically facilitate learning and synaptic plasticity (Sarantis et al., 2009; Scott and Aperia 2009; Smith-Roe and Kelley, 2000). In line with this proposal, systemic administration of a D1 or NMDA receptor antagonist impairs the acquisition of sexual experience (Bialy et al., 2000, 2010).

Other structures of the mesolimbic system are involved in functions that might be related to sexual experience-induced facilitation of sexual behavior. The BLA modulates emotionally influenced memory consolidation (McGaugh, 2002; LaLumiere and McGaugh, 2005), the PFC is involved in working memory (Seamans and Yang, 2004) and the NAcc, together with the BLA and PFC, is important for modulating memories (LaLumiere et al., 2005) and integrates the information from these structures influencing motor/behavioral output via projections to the VTA and ventral pallidum (LaLumiere and Kalivas, 2007).

Processing and anticipation of reward have been shown to affect firing patterns of hippocampal neurons (Tamura et al., 1992); this is particularly relevant for our understanding of motivated behavior because hippocampal glutamatergic afferents to the NAcc have the ability to regulate the activity of medium spiny neurons (O'Donnell and Grace, 1995).

On the other side, the participation of cognitive functions in the improvement of sexual behavior expression induced by sexual experience has also been postulated. According to this view, a representation of the potential sexual partner activates conditioned sexual behavior and the sexually conditioned stimuli modulate the effectiveness of the sexually associated signs provided by a sexual partner (Domjan, 2002). The author of this hypothesis concludes that sexual behavior is not the automatic outcome of the exposure to the sexual stimuli provided by potential sexual partners, but that the effectiveness of these stimuli are regulated by contextual cues, temporal factors, and learning and memory of previous sexual experiences.

Conclusions

The evidence reviewed in this chapter sustains the statement that instinctive behaviors are not only the expression of developmentally programmed neural circuits without the participation of the environment. The capacity of external cues to finely shape innate behaviors through learning and memory appears to be also relevant for instinctive behaviors. The robustness and stability of the genetically programmed behavioral sequences appear to be strengthened by experience. The mechanisms

involved in experience-induced brain adaptations are a matter of current investigation. Brain plasticity phenomena at the mesocorticolimbic system, triggered by reward-related learning, seem to play an important role in the adaptations that respond to the environmental demands.
Behavioral adaptations are essential for survival, even for instinctive behaviors.

References

Ågmo, A., Berenfeld, R., 1990. Reinforcing properties of ejaculation in the male rat: role of opioids and dopamine. Behav. Neurosci. 104, 177–182.

Alcock, J., 2005. Animal Behavior: An Evolutionary Approach. eighth ed. Sinauer Associates, Sunderland, MA, p. 564.

Arendash, G.W., Gorski, R.A., 1983. Effects of discrete lesions of the sexually dimorphic nucleus of the preoptic area or other medial preoptic regions on the sexual behavior of male rats. Brain Res. Bull. 10, 147–154.

Ballard, C.L., Wood, R.I., 2007. Partner preference in male hamsters: steroids, sexual experience and chemosensory cues. Physiol. Behav. 91, 1–8.

Barnard, C.J., 2004. Animal Behavior: Mechanisms Development, Function and Evolution. Rearson Education Ltd, Essex, UK, p. 726.

Bermant, G., Taylor, L., 1969. Interactive effects of experience and olfactory bulb lesions in male rat copulation. Physiol. Behav. 4, 13–17.

Berridge, K.C., Robinson, T.E., 1998. What is the role of dopamine in reward: hedonic impact, reward learning, or incentive salience? Brain Res. Rev. 28, 309–369.

Bialy, M., Rydz, M., Kaczmarek, L., 2000. Precontact 50-kHz vocalizations in male rats during acquisition of sexual experience. Behav. Neurosci. 114, 983–990.

Bialy, M., Kalata, U., Nikolaev-Diak, A., Nikolaev, E., 2010. D1 receptors involved in the acquisition of sexual experience in male rats. Behav. Brain Res. 206, 166–176.

Bloch, G.J., Davidson, J.M., 1968. Effects of adrenalectomy and experience on postcastration sex behavior in the male rat. Physiol. Behav. 3, 461–465.

Bolhuis, J.J., Giraldeau, L.-A., 2005. The Behavior of Animals: Mechanisms, Function, and Evolution. Blackwell Publishing, Malden, MA, USA, p. 515.

Bradley, K.C., Boulware, M.B., Jiang, H., Doerge, R.W., Meisel, R.L., Mermelstein, P.G., 2005. Changes in gene expression within the nucleus accumbens and striatum following sexual experience. Gene. Brain Behav. 4, 31–44.

Burkhardt, R.W., 2005. Patterns of Behavior: Konrad Lorenz, Niko Tinbergen, and the Founding of Ethology. University of Chicago Press, Chicago, IL, USA, p. 609.

Claro, F., Segovia, S., Guilamon, A., Del Abril, A., 1995. Lesions in the medial posterior region of the BST impair sexual behavior in sexually experienced and inexperienced male rats. Brain Res. Bull 36, 1–10.

Constantini, R.M., Park, J.H., Beery, A.K., Paul, M.J., Ko, J.J., Zucker, I., 2007. Postcastration retention of reproductive behavior and olfactory preferences in male Siberian hamsters: role of prior experience. Horm. Behav. 51, 149–155.

Coria-Avila, G.A., Pfaus, J.G., 2007. Neuronal activation by stimuli that predict sexual reward in female rats. Neuroscience. 148, 623–632.

Crawford, L.L., Holloway, K.S., Domjan, M., 1993. The nature of sexual reinforcement. J. Exp. Anal. Behav. 60, 50–66.

Damsma, G., Pfaus, J.G., Wenkstern, D., Phillips, A.G., Fibiger, H.C., 1992. Sexual behavior increases dopamine transmission in the nucleus accumbens and striatum of male rats: comparison with novelty and locomotion. Behav. Neurosci. 106, 181−191.

de Jonge, F.H., Louwerse, A.L., Ooms, M.P., Evers, P., Endert, E., van de Poll, N.E., 1989. Lesions of the SDN−POA inhibit sexual behavior of male Wistar rats. Brain Res. Bull. 23, 483−492.

Dewsbury, D.A., 1969. Copulatory behaviour of rats (Rattus norvegicus) as a function of prior copulatory experience. Anim Behav. 17 (2), 217−223.

Dewsbury, D.A., 1979. Description of sexual behavior in research on hormone-behavior interactions. In: Beyer, C. (Ed.), Endocrine Control of Sexual Behavior. Raven Press, New York, NY, pp. 3−32.

Domjan, M., 2002. Cognitive modulation of sexual behavior. In: Bekoff, M., Allen, C., Burghardt, G.M. (Eds.), The Cognitive Animal: Empirical and Theoretical Perspectives on Animal Cognition. The MIT Press, Cambridge, MA; London, UK, pp. 89−95.

Dugger, B.N., Morris, J.A., Jordan, C.L., Breedlove, S.M., 2008. Gonadal steroids regulate neural plasticity in the sexually dimorphic nucleus of the preoptic area of adult male and female rats. Neuroendocrinology 88, 17−24.

Edinger, K.L., Frye, C.A., 2007. Sexual experience of male rats influences anxiety-like behavior and androgen levels. Physiol. Behav. 92, 443−453.

El-Ghundi, M., O'Dowd, B.F., George, S.R., 2007. Insights into the role of dopamine receptor systems in learning and memory. Rev. Neurosci. 18, 37−66.

Fiorino, D.F., Kolb, B.S., 2003. Sexual Experience Leads to Long-Lasting Morphological Changes in Male Rat Prefrontal Cortex, Parietal Cortex, and Nucleus Accumbens Neurons. Society for Neuroscience, New Orleans, LA, 2003 Abstract Viewer and Itinerary Planner Washington, DC.

Fiorino, D.F., Coury, A., Phillips, A.G., 1997. Dynamic changes in nucleus accumbens dopamine efflux during the Coolidge effect in male rats. J. Neurosci. 17, 4849−4855.

Hedges, V.L., Chakravarty, S., Nestler, E.J., Meisel, R.L., 2009. Delta FosB overexpression in the nucleus accumbens enhances sexual reward in female Syrian hamsters. Genes Brain Behav. 8 (4), 442−449.

Hosokawa, N., Chiba, A., 2005. Effects of sexual experience on conspecific odor preference and estrous odor-induced activation of the vomeronasal projection pathway and the nucleus accumbens in male rats. Brain Res. 1066, 101−108.

Hull, E.M., Dominguez, J.M., 2007. Sexual behavior in male rodents. Horm. Behav. 52, 45−55.

Hull, E.M., Rodríguez-Manzo, G., 2009. Male sexual behavior. In: second ed. Pfaff, D.W., Arnold, A.P., Etgen, A.M., Fahrbach, S.E., Rubin, R.T. (Eds.), Hormones, Brain and Behavior, vol. 1. Academic Press, San Diego, CA, pp. 5−65.

Jay, T.M., 2003. Dopamine: a potential substrate for synaptic plasticity and memory mechanisms. Prog. Neurobiol. 69, 375−390.

Kandel, E.R., Squire, L.R., 2009. Memory: From Mind to Molecules. Scientific Library, New York, NY, p. 235.

Kippin, T.E., Pfaus, J.G., 2001. The development of olfactory conditioned ejaculatory preferences in the male rat. Nature of the unconditioned stimulus. Physiol. Behav. 73, 457−469.

Kohlert, J.G., Meisel, R.L., 1999. Sexual experience sensitizes mating-related nucleus accumbens dopamine responses of female Syrian hamsters. Behav. Brain Res. 99, 45−52.

Lagoda, G., Muschamp, J.W., Vigdorchik, A., Hull, E.M., 2004. A nitric oxide synthesis inhibitor in the medial preoptic area inhibits copulation and stimulus sensitization in male rats. Behav. Neurosci. 118, 1317−1323.

LaLumiere, Kalivas, 2007. Reward and drugs of abuse. In: Kesner, R.P., Martinez, J.L. (Eds.), Neurobiology of Learning and Memory, second ed. Academic Press, San Diego, CA, USA, pp. 459–482.

LaLumiere, R.T., McGaugh, J.L., 2005. Memory enhancement induced by posttraining intrabasolateral amygdala infusions of beta-adrenergic or muscarinic agonists requires activation of dopamine receptors: involvement of right, but not left, basolateral amygdala. Learn. Mem 12, 527–532.

LaLumiere, R.T., Nawar, E.M., McGaugh, J.L., 2005. Modulation of memory consolidation by the basolateral amygdala or nucleus accumbens shell requires concurrent dopamine receptor activation in both brain regions. Learn Mem. 12, 296–301.

Larsson, K., 1956. Conditioning and sexual behavior in the male albino rat. Almqvist and Wiksell, Stockholm, Sweden, 269 pp.

Larsson, K., 1967. Testicular hormone and developmental changes in mating behavior of the male rat. J. Comp. Physiol. Psychol. 63, 223–230.

Leuner, B., Glasper, E.R., Gould, E., 2010. Sexual experience promotes adult neurogenesis in the hippocampus despite an initial elevation in stress hormones. PLoS ONE 5, e11597. 10.1371/journal.pone.0011597.

Lopez, H.H., Ettenberg, A., 2002. Exposure to female rats produces differences in c-fos induction between sexually naïve and experienced male rats. Brain Res. 947, 57–66.

Lumley, L.A., Hull, E.M., 1999. Effects of a D-1 antagonist and of sexual experience on copulation-induced Fos-like immunoreactivity in the medial preoptic nucleus. Brain Res. 829, 55–68.

Manning, A., Thompson, M.L., 1976. Postcastration retention of sexual behavior in the male BDF1 mouse: the role of experience. Anim. Behav. 24, 523–533.

Markowski, V.P., Eaton, R.C., Lumley, L.A., Moses, J., Hull, E.M., 1994. A D1 agonist in the MPOA facilitates copulation in male rats. Pharmacol. Biochem. Behav. 47, 483–486.

McGaugh, J.L., 2002. Memory consolidation and the amygdala: a systems perspective. Trends Neurosci. 25, 456–461.

McHenry, J.A., Bell, G.A., Parrish, B.B., Hull, E.M., 2012. Dopamine D1 receptors and phosphorylation of dopamine- and cyclic AMP-regulated phosphoprotein-32 in the medial preoptic area are involved in experience-induced enhancement of male sexual behavior in rats. Behav. Neurosci. 126, 523–529.

Meredith, M., 1986. Vomeronasal organ removal before sexual experience impairs male hamster mating behavior. Physiol. Behav. 36, 737–743.

Mitchell, J.B., Gratton, A., 1994. Involvement of mesolimbic dopamine neurons in sexual behaviors: implications for the neurobiology of motivation. Rev. Neurosci. 5, 317–329.

O'Donnell, P., Grace, A.A., 1995. Synaptic interactions among excitatory afferents to nucleus accumbens neurons: hippocampal gating of prefrontal cortical input. J. Neurosci. 15, 3622–3639.

Olsen, C.M., 2011. Natural rewards, neuroplasticity, and non-drug addictions. Neuropharmacology 61, 1109–1122.

Pfaus, J.G., Phillips, A.G., 1991. Role of dopamine in anticipatory and consummatory aspects of sexual behavior in the male rat. Behav. Neurosci. 105, 725–741.

Pfaus, J.G., Wilkins, M.F., 1995. A novel environment disrupts copulation in sexually naïve but not experienced male rats: reversal with naloxone. Physiol. Behav. 57, 1045–1049.

Pfaus, J.G., Damsma, G., Nomikos, G.G., Wenkstern, D.G., Blaha, C.D., Phillips, A.G., et al., 1990. Sexual behavior enhances central dopamine transmission in the male rat. Brain Res. 530, 345–348.

Pfaus, J.G., Kippin, T.E., Centeno, S., 2001. Conditioning and sexual behavior: a review. Horm. Behav. 40, 291−321.

Pitchers, K., Balfour, M.E., Lehman, M.N., Richtand, N.M., Yu, L., Coolen, L.M., 2010a. Neuroplasticity in the mesolimbic system induced by natural reward and subsequent reward abstinence. Biol. Psych. 67, 872−879.

Pitchers, K., Frohmader, K., Vialou, V., Mouzon, E., Nestler, E., Lehman, M., et al., 2010b. DeltaFosB in the nucleus accumbens is critical for reinforcing effects of sexual reward. Gene. Brain Behav. 9, 831−840.

Pitchers, K.K., Schmid, S., Di Sebastiano, A.R., Wang, X., Laviolette, S.R., Lehman, M.N., et al., 2012. Natural reward experience alters AMPA and NMDA receptor distribution and function in the nucleus accumbens. PLoS ONE 7, e34700.

Powell, W.S., Dominguez, J.M., Hull, E.M., 2003. An NMDA antagonist impairs copulation and the experience-induced enhancement of male sexual behavior in the rat. Behav. Neurosci. 117, 69.

Robertson, G.S., Pfaus, J.G., Atkinson, L.J., Matsumura, H., Phillips, A.G., Fibiger, H.C., 1991. Sexual behavior increases c-fos expression in the forebrain of the male rat. Brain Res. 15, 352−357.

Rodríguez-Manzo, G., Guadarrama-Bazante, I.L., Morales-Calderón, A., 2011. Recovery from sexual exhaustion-induced copulatory inhibition and drug hypersensitivity follow a same time course: two expressions of a same process? Behav. Brain Res. 217, 253−260.

Saito, T.R., Moltz, H., 1986. Copulatory behavior of sexually naïve and sexually experienced male rats following removal of the vomeronasal organ. Physiol. Behav. 37, 507−510.

Sarantis, K., Matsokis, N., Angelatou, F., 2009. Synergistic interactions of dopamine D1 and glutamate NMDA receptors in rat hippocampus and prefrontal cortex: involvement of ERK1/2 signaling. Neuroscience 163, 1135−1145.

Schultz, W., 2006. Behavioral theories and the neurophysiology of reward. Ann. Rev. Psychol. 57, 87−115.

Schulz, K.M., Molenda-Figueira, H.A., Sisk, C.L., 2009. Back to future: the organizational-activational hypothesis adapted to puberty and adolescence. Horm. Behav. 55, 597−604.

Scott, L., Aperia, A., 2009. Interaction between N-methyl-D-aspartic acid receptors and D1 dopamine receptors: an important mechanism for brain plasticity. Neuroscience 158, 62−66.

Seamans, J.K., Yang, C.R., 2004. The principal features and mechanisms of dopamine modulation in the prefrontal cortex. Prog. Neurobiol. 74, 1−58.

Simerly, R.B., 2005. Wired on hormones: endocrine regulation of hypothalamic development. Curr. Opin. Neurobiol. 15, 81−85.

Smith-Roe, S.L., Kelley, A.E., 2000. Coincident activation of NMDA and dopamine D1 receptors within the nucleus accumbens core is required for appetitive instrumental learning. J. Neurosci. 20, 7737−7742.

Swaney, W.T., Curley, J.P., Champagne, F.A., Keverne, E.B., 2007. Genomic imprinting mediates sexual experience-dependent olfactory learning in male mice. Proc. Natl. Acad. Sci. USA. 104, 6084−6089.

Tamura, R., Ono, T., Fukuda, M., Nishijo, H., 1992. Monkey hippocampal neuron responses to complex sensory stimulation during object discrimination. Hippocampus. 2, 287−306.

Tenk, C.M., Wilson, H., Zhang, Q., Pitchers, K.K., Coolen, L.M., 2009. Sexual reward in male rats: effects of sexual experience on conditioned place preferences associated with ejaculation and intromissions. Horm. Behav. 55, 93−97.

Vanderwolf, C.H., 2003. An Odyssey Through the Brain, Behavior, and the Mind. Kluwer Academic Publishers, Boston, MA.

Vigdorchik, A.V., Parrish, B.P., Lagoda, G.A., McHenry, J.A., Hull, E.M., 2012. An NMDA antagonist in the MPOA impairs copulation and stimulus sensitization in male rats. Behav. Neurosci. 126, 186–195.

Wise, R.A., Spindler, J., deWit, H., Gerber, G.J., 1978. Neuroleptic-induced "anhedonia" in rats: pimozide blocks reward quality of food. Science 201, 262–264.

Woodson, J.C., 2002. Including 'learned sexuality' in the organization of sexual behavior. Neurosci. Biobehav. Rev. 26, 69–80.

Yao, W.D., Spealman, R.D., Zhang, J., 2008. Dopaminergic signaling in dendritic spines. Biochem. Pharmacol. 75, 2055–2069.

Zellner, M., Ranaldi, R., 2010. How conditioned stimuli acquire the ability to activate VTA dopamine cells: a proposed neurobiological component of reward-related learning. Neurosci. Biobehav. Rev. 34, 769–780.

Zellner, M.R., Kest, K., Ranaldi, R., 2009. NMDA receptor antagonism in the ventral tegmental area impairs acquisition of reward-related learning. Behav. Brain Res. 197, 442–449.

10 Memory Disorders: The Diabetes Case

Gustavo Liy-Salmeron and Shely Azrad

Facultad de Ciencias de la Salud, Universidad Anáhuac México Norte, Huixquilucan Edo. de México

Introduction

Diabetes mellitus is a common disorder affecting about 5% of the population worldwide (Whitmer, 2007) and 20% of people over 65 years (Arvanitakis et al., 2004). Type 2 diabetes (T2DM) is a heterogeneous metabolic disorder, characterized by reduced insulin sensitivity and relative insulin deficiency. Coexisting disorders, including obesity, hypertension, and dyslipidemia, contribute to the severity of T2DM. By contrast with type 1 diabetes, macrovascular disease causes about 80% of mortality in people with T2DM (McCrimmon et al., 2012). In addition to various adverse health defects, diabetic conditions are associated with modified brain function, namely with cognitive deficits. Diabetes is associated with a 50–100% increased risk of Alzheimer's disease (AD) and a 100–150% increased risk of vascular dementia (McCrimmon et al., 2012).

Reduced peripheral glucose regulation and diabetic conditions affect the central nervous system (CNS) through largely undetermined processes, contributing to diabetic encephalopathy. In particular, diabetic individuals display an augmented incidence of cognitive problems, which are particularly associated with atrophy of the hippocampal formation, which is involved in learning and memory processing. Probably because patients with diabetes have significantly lower cerebral blood flow than do healthy controls (McCrimmon et al., 2012; Duarte et al., 2012). Diabetes and prediabetes accelerated the progression from mild cognitive impairment to dementia by 3.18 years (Xu et al., 2010).

Findings from brain imaging and neuropathologic studies support the notion that the increased risk of cognitive decline and dementia in elderly people with diabetes reflects a dual pathologic process involving both cerebrovascular damage and neurodegenerative changes. In addition to vascular pathways, several possible pathophysiologic mechanisms, including hyperglycemia, insulin resistance, oxidative stress, advanced glycation end products, and inflammatory cytokines, may explain the effect of glucose deregulation on dementia risk. Recent genetic studies have found that chromosome 10 contains the genes for both late-onset (older than

Identification of Neural Markers Accompanying Memory. DOI: http://dx.doi.org/10.1016/B978-0-12-408139-0.00010-9

65 years) AD and T2DM (Martins et al., 2006). The first study to show clear evidence of hippocampal damage (volume losses and impairments in memory and learning) in T2DM shows that middle-aged individuals with well-controlled T2DM have clear deficits in hippocampal-based (recent or declarative) memory and selective Magnetic resonance imaging-based atrophy of the hippocampus relative to matched control subjects (Gold et al., 2007).

Glucose Regulation, Diabetes, and AD

Glucose metabolism and brain function in AD studies using positron emission tomography (PET) have consistently documented decreased brain glucose metabolism in moderately and severely demented patients compared to age-matched normal individuals (Ishii et al., 2001), and recent studies have shown impaired glucose utilization in neocortical association areas of the brains of patients with mild cognitive deficit, a precursor to AD (Nordberg et al., 2013).

In diabetic individuals, therefore, impairments in glucose and insulin regulation may contribute to AD pathology through mechanisms including decreased cortical glucose utilization, particularly in the hippocampus and entorhinal cortex. Studies have in fact found a correlation between blood glucose levels and memory performance in AD patients (Craft et al., 1992), although in the fasting state, others have observed no differences in resting glucose and insulin levels in AD patients compared to age-matched controls (Hassing et al., 2002).

Myo-inositol is located in astrocytes and its concentration changes in brain disease. It is increased in the frontal white matter of elderly people with T2DM. Concentrations of myo-inositol correlate with the presence of macrovascular disease and complications, but not with Glycated hemoglobin, suggesting that frontal gliosis can arise secondary to cerebrovascular changes (Ajilore et al., 2007; Geissler et al., 2003).

While glucose uptake in peripheral tissues requires insulin, in the brain this is considered to be an insulin-independent process. Insulin and its receptors are present in many parts of the brain. They are most dense in the hippocampus, hypothalamus, olfactory bulbs, and limbic system, whereas insulin itself is in highest concentration in the olfactory bulb and hypothalamus. Concentrations of insulin receptors (IRs) in the brain are particularly high in neurons, with many IR proteins in both cell bodies and synapses.

Insulin, IR, and AD

IRs are present in one of two isoforms; the IR-A isoform that lacks exon 11 that the other isoform, IR-B, expresses. A major functional difference between the two isoforms is that IR-A has a higher affinity for the neurotrophic factor insulin-like growth factor 2 (IGF-II) and a slightly higher affinity for insulin and has also been shown to associate/ dissociate with insulin quicker than IR-B. Brain-specific IRs are mainly the IR-A isoform and have a lower molecular weight than their peripheral counterparts (Martins et al., 2006).

There is a hypothesis that the concomitant loss of insulin and IGFs is the dominant cause for age-dependent, progressive brain atrophy with degeneration and cognitive decline. These tests are the first to show that insulin and IGFs regulate adult brain mass by maintaining brain protein content. Insulin and IGF levels are reduced in diabetes, and replacement of both ligands can prevent loss of total brain protein, widespread cell degeneration, and demyelination. IGF alone prevents retinal degeneration in diabetic rats. It supports synapses and is required for learning and memory. Replacement doses in diabetic rats can cross the blood−brain barrier to prevent hippocampus-dependent memory impairment (Serbedžija and Ishii, 2012).

Interestingly, IR activation can also mediate vasoconstriction. Activation of IR can also lead to phosphorylation of Shc which then binds Growth factor receptor-bound protein 2 resulting in activation of Sos. This complex then activates Ras leading to phosphorylation Raf which results in activation of Mitogen-activated protein kinases. Activation of MAPK stimulates release of endothelin-1, a vasoconstrictor. By mediating vascular properties, insulin signaling plays a significant role in glucose and oxygen availability to the brain. Conversely, dysfunction in insulin signaling, as observed in T2DM, has profound detrimental effects on hemodynamics and, thus, maintenance of normative brain function (Formoso et al., 2006).

Insulin signaling is known to protect against oxidative stress, mitochondrial collapse, over-activity of Glycogen synthase kinase 3 β leading to hyperphosphorylation of tau, activation of death promoting transcription factors, and formation of apoptotic structures. Loss of insulin signaling in the brain leaves neurons vulnerable to a myriad of insults (Reddy, 2013).

Although it is known that IRs and insulin itself are present in the brain, the exact physiologic purpose in the brain is unclear. Insulin may also function as a neuromodulator directing the secretion and reuptake of neurotransmitters and affecting learning and memory.

Insulin and/or IRs appear to contribute to learning and memory via the activation of specific signaling pathways, one of which is associated with long-term memory formation; therefore, desensitization of the neuronal IR, which occurs in diabetes and the insulin-resistance syndrome, may be another key factor in the pathogenesis of AD. Insulin levels may also have an impact on the regulation of Ab proteolytic degradation, as the insulin degrading enzyme (IDE) can break down several peptides, including insulin, Ab, glucagon, and amylin (Martins et al., 2006; Formoso et al., 2006).

Apolipoprotein E and AD

Fifty percent of the early-onset familial AD is caused by autosomal dominant inheritance of mutations in the amyloid precursor protein (APP) or presenilin genes but the majority of AD cases are late-onset, and are not thought to be due to genetic mutations. The risk factors for AD are high cholesterol levels, obesity, diabetes, coronary artery disease, polymorphisms in the gene that codes for low-density lipoprotein receptor-related protein-1 (LRP-1), and the possession of one or more apolipoprotein E e4 (APOE e4) alleles and this are all physiologically or biochemically connected: they are all associated in some way with the transport and metabolism of lipids. In the periphery, APOE aids the transport of triglyceride, phospholipid, cholesteryl esters, and cholesterol into cells, by mediating the binding, internalization, and catabolism of lipoprotein particles. It is the main ligand for the Low-density lipoprotein receptor found on the liver and other tissues, and for the specific apoE receptor (chylomicron remnant) of hepatic tissues. In human cerebrospinal fluid, most of the apolipoprotein content is represented by apoE and apoA and these are present on astrocyte-secreted lipoproteins which have a density similar to plasma High-density lipoprotein. ApoE is required for lipoprotein uptake via LDL receptors and LRP-1 in the CNS, most likely in order to mediate the uptake and redistribution of lipids and cholesterol within the CNS, as it does in the periphery (Hertze et al., 2013).

AD is characterized histologically by the presence of intracellular and extracellular amyloid deposits in the brain, together with widespread neuronal cell loss. Extracellular amyloid deposits are known as neuritic or senile plaques. The main protein constituent of AD senile plaques, a peptide known as amyloid beta (Aβ) is a normal proteolytic product of a much larger transmembrane protein, the amyloid precursor protein (Tsitsopoulos and Marklund, 2013). In AD, Aβ peptides aggregate into insoluble fibrils which deposit in the brain to produce the characteristic amyloid plaques. Aβ oligomers are known impair insulin signaling in neurons by competing with insulin for receptor binding sites and studies have linked Aβ oligomers to decreased IR numbers (Duarte et al., 2012). Aβ can be detected in plasma, cerebrospinal fluid and in cell culture media. Brain pathology from type II diabetic patients frequently includes amyloid deposition (Martins et al., 2006; Tsitsopoulos and Marklund, 2013).

Aβ appears to be a direct competitive inhibitor of insulin binding to its receptor. One of the many effects of insulin binding to the IR is the promotion of non-amyloidogenic soluble APP (sAPP) secretion, and Ab binding to IRs can inhibit this effect. Inhibitor studies have implicated the phosphatidyl inositol 3 kinase (PI3K) signaling pathway in the promotion of sAPP secretion by insulin (Martins et al., 2006).

Cholesterol and AD

In AD patients, the cholesterol flux is elevated. The adenosine triphosphate-binding cassette transporter (ABCA1) is a major regulator of plasma HDL: it transports cellular cholesterol and phospholipids from cells onto HDL. ABCA1 plays a rate-limiting role in the process by which peripheral cholesterol is transported back to the liver for metabolism and subsequent excretion. It has been hypothesized that if ABCA1 also stimulates cholesterol flux from the CNS to the periphery, it is likely to cause an increase in the internal cycling of brain cholesterol, and thereby prevent the accumulation of excess cholesterol in neurons. In fact, ABCA1 has been detected in neurons, and increased expression is accompanied by cholesterol efflux from neurons and glia. Increased neuronal expression of ABCA1 also affects APP processing, causing a decrease in Aβ production (Martins et al., 2006; Tsitsopoulos and Marklund, 2013). Individuals with a genetic polymorphism (R219K) in the ABCA1 gene have 30% lower cholesterol in their cerebrospinal fluid, this polymorphism is therefore likely to modify brain cholesterol metabolism. Interestingly, this polymorphism is associated with a 1.7 year delay in AD age of onset. Activation of liver X receptors (LXRs) has been shown to stimulate ABCA1 levels and decrease Ab concentrations (Sun et al., 2003). Therefore, LXR activation may provide a novel approach for the treatment of AD.

Higher levels of plasma glucose cause advanced glycation endproducts (AGEs) formation in diabetic patients. Accumulation of AGEs in those with AD is thought to be due to accelerated oxidation of glycated protein. Aβ plaques, proteins present in AD, are a precursor for the formation of AGEs. AGE formation has several metabolic sequelae including oxidative stress, glucose hypometabolism, and impaired cell function. AGEs impair neuronal functioning through a variety of mechanisms including apoptosis, calcium influx, oxidative stress, and inhibition of oxidative phophorylation. AGEs can affect neuronal function by modifying functionally important proteins. In tissues most affected by diabetic complications, AGEs are present in highest concentrations, resulting in structural changes in cell membranes and intracellular components (Yamagishi et al., 2005).

Animal models of induced diabetes suggest a direct neurodegenerative effect of diabetes. Most studies show results in the hippocampus, the major area associated with learning and memory because it is more susceptible than other brain regions to damage by all sorts of insults, including severe hypoglycemia and hypoxia; thus, it is plausible that the hippocampus is the first region to be affected by T2DM (Gold et al., 2007; Mastrocola et al., 2012). These results suggest that diabetes is associated (i) with changes in hippocampal synaptic plasticity that is related to the degree of hyperglycemia but is reversible by glycemic control; (ii) with molecular changes in hippocampal neurons, including damage to presynaptic and postsynaptic structures; and (iii) in long-term diabetes, with decreased neuronal densities in the CA-1 region on the hippocampus. The hippocampus is also the first structure to be affected by the neurodegeneration of AD. Maybe this is why the risk of incident AD is 65% higher in those with diabetes than in those without it.

Researchers have found that the administration of glucose (50 g) to moderately to severely demented patients (probable AD) did improve memory. Deficits in glucose metabolism might also potentiate the neuronal cell death produced by other pathological processes (such as cerebral hypoperfusion, abnormal cholesterol metabolism, or high levels of toxic Aβ), which in turn might be influenced by genetic predisposition such as possession of APOE e4 alleles (Martins et al., 2006).

Recent clinical studies have shown that induced hyperinsulinemia causes an increase in plasma and brain Ab1β42 levels, as well as increased levels of inflammation markers in cerebrospinal fluid. Insulin resistance also leads to a functional decrease in IR-mediated signal transduction in the brain.

Caffeine and AD

An emerging candidate to manage diabetes-induced neurodegeneration is caffeine. Caffeine is the most widely consumed psychoactive substance and acts as an antagonist of adenosine A1 receptors (A1R) and A2AR at nontoxic doses. Caffeine consumption alleviates cognitive impairment in both humans and animals, namely in AD and affords protection upon CNS injury. Furthermore, several studies indicate that habitual coffee consumption reduces the risk of diabetes. It has been found that streptozotocin-induced diabetes modifies the expression and density of adenosine receptors in the hippocampus (Ponce-Lopez et al., 2011), as occurs in most noxious brain conditions and caffeine prevents streptozotocin-induced neurotoxicity.

Caffeine consumption restored memory performance and abrogated the diabetes-induced loss of nerve terminals and astrogliosis. These results provide the first evidence that type 2 diabetic mice display a loss of nerve terminal markers and astrogliosis, which is associated with memory impairment; furthermore, caffeine consumption prevents synaptic dysfunction and astrogliosis as well as memory impairment in T2DM (Duarte et al., 2012).

Conclusions

There is now substantial evidence to support the conclusion that T2DM is associated with memory impairment and dementia. These alterations in cognitive function may represent accelerated aging in diabetes, although evidence in this regard is poor. There is literature suggesting that both the metabolic control and the hypertension that occurs as a consequence of diabetes are involved in the etiology of the cognitive impairment that may occur. The literature on the mechanisms with regard to cognitive impairment is unclear. Evidence suggests a role for glycemic control in both memory decline and dementia. The role of insulin, hypertension, and dyslipidemia in loss of memory functions is less convincing, although these factors are associated with decline in other functions such as verbal fluency. Their role in dementia is also apparent. It is unclear if these factors are using the same or different mechanisms to cause these changes in cognition. It is possible that different pathways may converge in the case of dementia. More research is needed to further characterize these mechanisms. The association between factors associated with diabetes and cognitive function as well as dementia provides the individual with means to control the onset of these consequences of aging and pathology by modification of lifestyle and behavior.

References

Ajilore, O., Haroon, E., Kumaran, S., et al., 2007. Measurement of brain metabolites in patients with type 2 diabetesand major depression using proton magnetic resonance spectroscopy. Neuropsychopharmacology. 32, 1224−1231.

Arvanitakis, Z., Wilson, R.S., Bienias, J.L., Evans, D.A., Bennett, D.A., 2004. Diabetes mellitus and risk of Alzheimer disease and decline in cognitive function. Arch. Neurol. 61 (5), 661−666.

Craft, S., Zallen, G., Baker, L.D., 1992. Glucose and memory in mild senile dementia of the Alzheimer type. J. Clin. Exp. Neuropsychol. 14 (2), 253−267.

Duarte, J.M., Agostinho, P.M., Carvalho, R.A., Cunha, R.A., 2012. Caffeine consumption prevents diabetes-induced memory impairment and synaptotoxicity in the hippocampus of NONcZNO10/LTJ mice. PLoS One 7 (4), e21899.

Formoso, G., Chen, H., Kim, J.A., Montagnani, M., Consoli, A., Quon, M.J., 2006. Dehydroepiandrosterone mimics acute actions of insulin to stimulate production of both nitric oxide and endothelin 1 via distinct phosphatidylinositol 3-kinase- and mitogen-activated protein kinase-dependent pathways in vascular endothelium. Mol. Endocrinol. 20 (5), 1153−1163.

Geissler, A., Fründ, R., Schölmerich, J., Feuerbach, S., Zietz, B., 2003. Alterations of cerebral metabolism in patients withdiabetes mellitus studied by proton magnetic resonance spectroscopy. Exp. Clin. Endocrinol. Diabetes 111, 421−427.

Gold, S.M., Dziobek, I., Sweat, V., Tirsi, A., Rogers, K., Bruehl, H., et al., 2007. Hippocampal damage and memory impairments as possible early brain complications of type 2 diabetes. Diabetologia 50 (4), 711−719.

Hassing, L.B., Johansson, B., Nilsson, S.E., Berg, S., Pedersen, N.L., Gatz, M., et al., 2002. Diabetes mellitus is a risk factor for vascular dementia, but not for Alzheimer's disease: a population-based study of the oldest old. Int. Psychogeriatr. 14 (3), 239−248.

Hertze, J., Palmqvist, S., Minthon, L., Hansson, O., 2013. Tau pathology and parietal white matter lesions have independent but synergistic effects on early development of Alzheimer's disease. Dement. Geriatr. Cogn. Dis. Extra. 3 (1), 113−122.

Ishii, K., Willoch, F., Minoshima, S., Drzezga, A., Ficaro, E.P., Cross, D.J., et al., 2001. Statistical brain mapping of 18F-FDG PET in Alzheimer's disease: validation of anatomic standardization for atrophied brains. J. Nucl. Med. 42 (4), 548−557.

Martins, I.J., Hone, E., Foster, J.K., Sünram-Lea, S.I., Gnjec, A., Fuller, S.J., et al., 2006. Apolipoprotein E, cholesterol metabolism, diabetes, and the convergence of risk factors for Alzheimer's disease and cardiovascular disease. Mol. Psychiatr. 11 (8), 721−736.

Mastrocola, R., Barutta, F., Pinach, S., Bruno, G., Perin, P.C., Gruden, G., 2012. Hippocampal heat shock protein 25 expression in streptozotocin-induced diabetic mice. Neuroscience 227, 154−162.

McCrimmon, R.J., Ryan, C.M., Frier, B.M., 2012. Diabetes and cognitive dysfunction. Lancet 379 (9833), 2291−2299.

Nordberg, A., Carter, S.F., Rinne, J., Drzezga, A., Brooks, D.J., Vandenberghe, R., et al., 2013. European multicentre PET study of fibrillar amyloid in Alzheimer's disease. Eur. J. Nucl. Med. Mol. Imaging 40 (1), 104−114.

Ponce-Lopez, T., Liy-Salmeron, G., Hong, E., Meneses, A., 2011. Lithium, phenserine, memantine and pioglitazone reverse memory deficit and restore phospho-GSK3β decreased in hippocampus in intracerebroventricular streptozotocin induced memory deficit model. Brain Res. 1426, 73−85.

Reddy, P.H., 2013. Amyloid beta-induced glycogen synthase kinase 3β phosphorylated VDAC1 in Alzheimer's disease: implications for synaptic dysfunction and neuronal damage. Biochim. Biophys. Acta pii: S0925-4439(13)00221-4.

Serbedžija, P., Ishii, D.N., 2012. Insulin and insulin-like growth factor prevent brain atrophy and cognitive impairment in diabetic rats. Indian J. Endocrinol. Metab. 16 (Suppl. 3), S601–S610.

Sun, Y., Yao, J., Kim, T.W., Tall, A.R., 2003. Expression of liver X receptor target genes decreases cellular amyloid beta peptide secretion. J. Biol. Chem. 278 (30), 27688–27694.

Tsitsopoulos, P.P., Marklund, N., 2013. Amyloid-β peptides and tau protein as biomarkers in cerebrospinal and interstitial fluid following traumatic brain injury: a review of experimental and clinical studies. Front Neurol. 4, 79.

Whitmer, R.A., 2007. Type 2 diabetes and risk of cognitive impairment and dementia. Curr. Neurol. Neurosci. Rep. 7 (5), 373–380.

Xu, W., Caracciolo, B., Wang, H.X., Winblad, B., Bäckman, L., Qiu, C., et al., 2010. Accelerated progression from mild cognitive impairment to dementia in people with diabetes. Diabetes 59 (11), 2928–2935.

Yamagishi, S., Nakamura, K., Inoue, H., Kikuchi, S., Takeuchi, M., 2005. Serum or cerebrospinal fluid levels of glyceraldehyde-derived advanced glycation end products (AGEs) may be a promising biomarker for early detection of Alzheimer's disease. Med. Hypothes. 64 (6), 1205–1207.

Printed in the United States
By Bookmasters